Lactose and Lactose Derivatives

Edited by Néstor Gutiérrez-Méndez

Published in London, United Kingdom

IntechOpen

Supporting open minds since 2005

Lactose and Lactose Derivatives
http://dx.doi.org/10.5772/intechopen.87477
Edited by Néstor Gutiérrez-Méndez

Contributors
José Manuel Pais-Chanfrau, Jimmy Núñez Pérez, Rosario del Carmen Espín-Valladares, Marcos Vinicio Lara-Fiallos, Luis Enrique Trujillo-Toledo, Antónia Macedo, Joana Monteiro, David Azedo, Elizabeth Duarte, Carlos Pereira, Lorena Mardones, Marcelo Villagrán, Sergio I. Martínez-Monteagudo, Kaavya Rathnakumar, Maryam Enteshari, Collette Nyuydze, Juan C. Osorio-Arias, Hiran Ranaweera, Néstor Gutiérrez-Méndez, José Carlos Rodríguez-Figueroa, Hada María Guevara-Alvarado, Esther Carrillo-Pérez, Einar Vargas-Bello-Pérez

First published in London, United Kingdom, 2020 by IntechOpen
IntechOpen is the global imprint of INTECHOPEN LIMITED, registered in England and Wales, registration number: 11086078, 5 Princes Gate Court, London, SW7 2QJ, United Kingdom
Printed in Croatia

British Library Cataloguing-in-Publication Data
A catalogue record for this book is available from the British Library

Additional hard and PDF copies can be obtained from orders@intechopen.com

Lactose and Lactose Derivatives
Edited by Néstor Gutiérrez-Méndez
p. cm.
Print ISBN 978-1-83962-604-3
Online ISBN 978-1-83962-605-0
eBook (PDF) ISBN 978-1-83962-606-7

For EU product safety concerns:
IN TECH d.o.o., Prolaz Marije Krucifikse Kozulić 3, 51000 Rijeka, Croatia, info@intechopen.com or visit our website at intechopen.com.

Meet the editor

Néstor Gutiérrez-Méndez is currently a full professor in the Chemistry Faculty at the Graduate School of Food Science, Autonomous University of Chihuahua, Mexico. He obtained his DSc from the Research Center for Food and Development (CIAD)-Mexico-Hermosillo in 2008. He is an active member of the National Researchers System (level 2) belonging to the Mexican Council of Science and Technology (CONACyT). He has authored/co-authored more than forty ISI journal articles and ten book chapters in prestigious editorials. His main research areas include dairy science and technology, the study of plant-derived compounds with antidiabetic effect, and the development of food-grade nano and microemulsions.

Contents

Preface

Lactose is a disaccharide with unique characteristics. For instance, the beta-glycosidic bond that joins galactose and glucose. This sugar also has a lower caloric value and a lower glycemic index than other mono and disaccharides. Furthermore, lactose sweetness is less than that of sucrose, and the crystals of lactose have excellent plasticity and compressibility properties. Consequently, lactose and lactose derivatives are widely used in the food and pharmaceutical industries. This book reviews some aspects of lactose properties and synthesis (Section 1) as well as the recent advances in the recovery of lactose and lactose derivatives from cheese whey (Section 2).

In the first section, two chapters describe lactose synthesis, its biological role in mammals, and its key physicochemical properties. The exclusive source of lactose is mammals' milk since this sugar is synthesized in the mammary glands by the Golgi apparatus of alveolar epithelial cells. Unlike other disaccharides, a beta-glycosidic bond joins the galactose and glucose in a lactose molecule. Accordingly, lactose metabolism requires a unique enzyme (lactase) able to hydrolyze the beta-glycosidic bond. Newborn and young mammals produce enough lactase enzyme to completely metabolize lactose; nevertheless, adult mammals metabolize lactose deficiently, leading to maldigestion problems upon consumption of milk or food products containing lactose. The food and pharmaceutical industries are aware of the lactose intolerance problems among the population. Hence, these industries are looking for lactose substitutes or derivatives that could be used in food and drug production.

In the second section, the first chapter provides information on recovering lactose from cheese whey through membrane technology. The following chapters are devoted to production of lactose derivatives, including lactitol, organic acids (acetic, ascorbic, butyric, citric, propionic, succinic, and lactic acid), lactulose, sialyl lactose, galacto-oligosaccharides, and lactosucrose. Cheese whey is a byproduct of the cheesemaking industry that contains chiefly water, lactose, proteins, and a minor proportion of minerals and fat. The conventional method of recovering lactose from cheese whey is crystallization, but membranes have recently gained attention in the dairy industry. Alternatively, the lactose in cheese whey can be transformed into organic acids by fermentation. Microorganisms like lactic acid bacteria are used to produce galacto-oligosaccharides or other polysaccharides. Lactose derivatives like lactitol are used in bakery, confectionery, chocolate, desserts, and chewing gum. This sugar alcohol is not found in nature, and it has a low caloric value (2.0 kcal g-1) since the lactase enzyme can barely hydrolyze lactitol. All these lactose derivatives open new perspectives in the use of lactose in the food and pharmaceutical industries.

I want to thank all the authors who contributed with their hard work and knowledge of lactose and lactose derivatives. This book would not have been possible without your excellent work. I also would like to thank IntechOpen Author Service Manager Ms. Sandra Maljavac for her support during the publication of this book.

Néstor Gutiérrez-Méndez
Chemistry Faculty,
Autonomous University of Chihuahua,
Chihuahua, Mexico

Section 1

Lactose Properties and Synthesis

Chapter 1

Introductory Chapter: Lactose

Néstor Gutiérrez-Méndez

1. The biological role of lactose

Milk provides infants with essential nutrients to support the first months of life. Newborns and young animals obtain their energy mostly from milk lipids and lactose (~17 kJ per gram of lactose). Only lactose provides 40% of the energy needs of suckling mammals. This fact explains why almost all the mammalian milk contains 40-75 g of lactose per litter, and why the milk of mammals is the only source in nature with a significant content of lactose [1–3]. Congenital deficiency to digest lactose is rare in baby mammals since it can lead to growth delay, dehydration, and even the death [3].

Lactose is a disaccharide synthesized in the mammary gland of mammals, and only scarce plant species show this saccharide. The Golgi vesicles of mammary epithelial cells synthesize lactose from two molecules of glucose. One of this glucose is first epimerized to galactose (Leloir pathway) and phosphorylated. Then, condensation with the other glucose occurs through the lactose synthetase system. This system comprises the enzyme galactosyl transferase and the protein modifier α-lactalbumin. When the protein modifier binds to the galactosyl transferase, it catalyzes the synthesis of lactose from uridine-diphosphate-galactose (UDP-gal) and glucose [1–3]. In the absence of the protein modifier, the galactosyl transferase does not synthesize lactose and instead catalyzes the synthesis of N-acetyl lactosamine on glycoproteins. This last reaction occurs in most tissues, but in the mammary gland of women after giving birth, the increase in prolactin and a decrease in progesterone hormones induce the formation of the protein modifier (α-lactalbumin). Consequently, the breast can synthesize lactose in the milk for the nourishment of newborn mammals [2, 3].

Lactose digestion in humans involves the action of intestinal lactase. Lactose is a disaccharide containing galactose and glucose linked by a β 1-4 glycosidic bond. This sugar cannot be transported across the epithelial cell membrane into the enterocytes and then into the bloodstream as a disaccharide. The release of galactose and glucose monomers by hydrolysis of the β-glycosidic bond allows their transport into the enterocytes through the Na^+ dependent transporter SGLT1. Then, the GLUT2 transporter carries these monosaccharides into the blood [2, 3]. The β-glycosidic bond in lactose molecules is hydrolyzed in the small intestine by a β-galactosidase. There are three types of these enzymes in human tissue: (a) the β-galactosidase in the lysosomes, (b) the β-galactosidase in the cytosol of cells, and (c) the β-galactosidase in the small intestine. It is worth to mention that the human intestinal β-galactosidase has similarities to intestinal lactases reported in rabbit (83%) and rat (77%). However, this enzyme has no sequence homology with the other two types of β-galactosidases in human tissue, the β-galactosidase in bacteria, or different kinds of β-galactosidases found in eukaryotic cells [2, 4].

The intestinal β-galactosidase (like other carbohydrate-hydrolyzing enzymes) is situated close to the brush border on the upper surface of enterocytes on the microvilli. Hundreds of tiny finger-like structures (villi) protrude from the small intestine wall. These villi have additional extensions (microvilli) that form the brush border of enterocytes. The small intestine has three segments, duodenum (5-6 cm long), jejunum (2.5 m long), and ileum (4-5 m long). The β-galactosidase is all over the small intestine, but primarily in the jejunum, where the pH is 7-8, and the bacterial concentration is low [2, 4]. The human intestinal β-galactosidase is unique because it has two different active sites within one polypeptide chain. Therefore, this enzyme can hydrolyze lactose, but also other types of substrates. One of the active sites hydrolyses lactose into galactose and glucose and cleaves other substrates like cellobiose, cellotriose, cellotetrose, and cellulose (EC 3.2.1.108). The other active site hydrolyses phlorizin, an aryl α-glucoside linked to phloretin (EC 3.2.1.62). The active area for phlorizin also cleavage β-glycosides with a sizeable hydrophobic chain like cerebrosides, made up of ceramide (sphingosine with a fatty acid attached) bonded by a β link to galactose or another hexose. The hydrolysis of cerebrosides provides sphingosine, a key molecule maintaining the membranes of the brain. Consequently, the full name of the intestinal β-galactosidase is lactase-phlorizin hydrolase or LPH [2–4].

2. Lactose intolerance

The loss of intestinal β-galactosidase (LPH) reduces humans' capability to metabolize lactose. The synthesis of LPH starts in humans during the gestation (8-34 weeks) and reaches its peak at birth. After the first 6-12 months of life, β-galactosidase begins to decline. Over the four years, at least 60% of people reduce their levels of LPH to 5-10%. The decline in lactase after weaning occurs in all mammals. [1–4]. The levels of LPH during adulthood vary significantly between ethnic groups. For instance, more than 90% of Chinese and Japanese adults have low lactase levels and potential lactose intolerance, in contrast with the only 10% of white Northern Europeans. The domestication of cattle by European populations promoted for centuries milk as a food item for adults. Therefore, many people in this ethnic group developed a persistent lactase expression during adulthood (lactase persistent) [2, 3].

Most adult individuals have reduced activity of LPH in the small intestine (lactase non-persistent, or wild-type condition), and only a minority of humans have a high level of LPH activity (lactase persistent). Any deficiency of intestinal β-galactosidase is considered hypolactasia, and a total lack of LPH activity in the small intestine is referred to as alactasia. This last condition is infrequent, and before the twentieth century, infants with congenital alactasia had a little expectation of surviving. There are two conditions for hypolactasia: primary adult hypolactasia (lactase non-persistent) and secondary adult hypolactasia (acquired hypolactasia). The primary hypolactasia is due to the normal decrement with the age of lactase quantity in the small intestine. This decrement occurs because the human body reduces the transcription of the lactase gen (LCT; NCBI reference sequence XP_016859577.1), or by a reduction in the translation of the mRNA. The reduction of lactase in adults does not mean automatically that these individuals will have problems digesting lactose. Some researchers estimate that 50% of the regular β-galactosidase activity is enough for adequate lactose assimilation. However, human adults with low quantities of LPH (<50%) in the small intestine cannot properly digest the lactose in 100 mL of milk (lactose maldigestion). The secondary hypolactasia is different; this derives from an intestinal infection (by bacteria, viruses, or protozoa), severe malnutrition, inflammatory bowel diseases, actinic enteritis, and extensive use of antibiotics (i.e., kanamycin, neomycin, polymycin, and tetracycline). Additionally, diverse substances in the gut lumen can induce inhibition of LPH activity. Nevertheless, secondary hypolactasia can be treatable and potentially reversible [1–4]. On the other hand, lactose intolerance occurs when a lactose maldigester shows gastrointestinal problems. Bacteria in the large intestine convert lactose into gases and diverse metabolites if this sugar is not hydrolyzed in the jejune by the LPH. The most common symptoms of lactose intolerance are the development of flatulence, abdominal distention, and diarrhea. There is no exact data, but about two-thirds of adult humans cannot digest lactose properly [2]. Therefore, nowadays, the dairy industry is looking to develop dairy products without lactose for consumers suffering from lactose intolerance.

3. Physicochemical properties of lactose

Lactose is a reducing disaccharide of galactose and glucose discovered in milk in the 17th century. Both the galactose and the glucose can form a hemiacetal link and create a ring structure. A β-glycosidic link connects the two pyranose structures deriving in a 4-O-β-D-galactopyranosyl-D-glucopyranose molecule. This disaccharide has a chiral center that exists as two isomers: α-lactose and β-lactose. The α-isomer rotates the plane of polarized light +92.6° and the β-isomer +34° at 20°C. When lactose is in an aqueous milieu, its ring structure opens and closes interchanging between the α- and β-isoforms (mutarotation). At some point, these isoforms acquire an equilibrium (mutarotation equilibrium). Lactose mutarotation is a slow process that is very temperature-dependent. For instance, at 18.8°C the mutarotation equilibrium is achieved in 6.5 hours with a proportion of 40% of α-lactose and 60% of β-lactose; but at 0°C, the stability can take up to 72 hours. Overall, the proportion of β-lactose is always higher than the α-lactose at mutarotation equilibrium, because the β-isoform is more soluble than the α-isoform. For example, at 35°C the solubility of α-lactose is 7 g per 100 g of water, in contrast, the solubility of β-lactose is 50 g per 100 g of water [2, 5–7]. Certainly, the solubility of both isoforms will decrease if the temperature drops. Like other sugars, lactose molecules nucleate and crystallize when the concentration of this sugar overcomes its maximum solubility at a specific temperature. The dairy industry applies this principle to crystallize lactose from whey, a by-product of cheesemaking [8].

This by-product of cheesemaking contains 0.8 – 1% protein, 0.06% fat, 4.5 – 6% lactose, and 90 – 92% of water. To crystallize lactose from the cheese whey, it needs to be first, defatted, deproteinated and evaporated to concentrate lactose between 39 and 56%. At this concentration, lactose will crystallize when the evaporated whey is cooled enough (i.e., 20-25°C). During the cooling step, lactose moves through and beyond the metastable zone (MZ), a region between the solubility and supersolubility of lactose. The spontaneous nucleation of lactose occurs when the supersolubility is exceeded, outside the MZ. Therefore, the width of the MZ determines the temperature drop necessary to induce lactose nucleation. After nucleation, crystals' growth depends on the degree of lactose saturation and the temperature, since the last one affects lactose solubility [7, 9–13]. The overall process of lactose crystallization is slow. In consequence, mutarotation can occur during the nucleation or the growth of crystals. However, if the mutarotation rate is lower than the crystallization rate, the kinetics of mutarotation will dominate over the nucleation and crystal growth. The industrial process of lactose crystallization from cheese whey is slow (up to 48 h) and requires an elevated lactose concentration to induce nucleation (high evaporation cost). Different approaches have been studied to overcome the drawbacks of lactose crystallization. Among these are the seeding of lactose nuclei, anti-solvents (i.e., ethanol and acetone), and the appliance of high-power ultrasound. Alternatively, methods other than crystallization have been investigated to recover lactose from the cheese whey, like the use of membranes [9, 11, 14–17].

4. Final remarks

Despite the persistence of lactose intolerance in the population, the dairy industry produces 400,000 tons of crystalline lactose worldwide [6]. The food and pharmaceutical industries use large amounts of lactose. Foods like instant coffee, infant formula, baked foods, and many others utilize lactose as an ingredient. This saccharide has a lower caloric value and a lower glycemic index than other carbohydrates. Additionally, lactose is less sweet than sucrose, and it has good plasticity and compressibility. These properties of lactose explain why most pharmaceutical pills contain lactose as a filling material. Derivative lactose compounds like lactic acid, lactitol, lactulose, and oligosaccharides gain interest in the food industry [5, 6, 12].

Author details

Néstor Gutiérrez-Méndez
Chemistry School, Autonomous University of Chihuahua, Chihuahua, Chihuahua, Mexico

*Address all correspondence to: ngutierrez@uach.mx

References

[1] Walstra, P., J.T.M. Wouters, and T.M. Geurst, Dairy Science and Technology. Second ed. 2006, Boca Raton FL: CRC press.

[2] Campbell, A.K., J.P. Waud, and S.B. Matthews, The Molecular basis of Lactose Intolerance. Science Progress, 2005. 88(3): p. 157-202.

[3] Fassio, F., M.S. Facioni, and F. Guagnini, Lactose Maldigestion, Malabsorption, and Intolerance: A Comprehensive Review with a Focus on Current Management and Future Perspectives. Nutrients, 2018. 10(11): p. 1599.

[4] Ingram, C.J.E. and D.M. Swallow, Lactose Malabsorption, in Advanced Dairy Chemistry: Volume 3: Lactose, Water, Salts and Minor Constituents, P. McSweeney and P.F. Fox, Editors. 2009, Springer New York: New York, NY. p. 203-229.

[5] Listiohadi, Y.D., Hourigan, J. A., Sleigh, R., Steele, R. J., Propierties of Lactose and its caking behaviour. Australian Journal of Dairy Technology, 2005. 60: p. 33-52.

[6] Fox, P.F., Lactose: chemistry and properties, in Advanced dairy chemistry volume 3: lactose, water, salts and minor constituents, P.F. Fox and P.L.H. McSweeney, Editors. 2009, Springer: New York. p. 1-15.

[7] Wong, S.Y. and R.W. Hartel, Crystallization in lactose refining-a review. J Food Sci, 2014. 79(3): p. R257-72.

[8] Zamanipoor, M.H. and R.L. Mancera, The emerging application of ultrasound in lactose crystallisation. Trends in Food Science & Technology, 2014. 38(1): p. 47-59.

[9] Sánchez-García, Y.I., Bhangu, S. K., Ashokkumar, M., & Gutiérrez-Méndez, N., Sonocrystallization of lactose from whey, in Technological Approaches for Novel Applications in Dairy Processing, INTECH, Editor. 2018. p. 51-71.

[10] de Castro, M.D. and F. Priego-Capote, Ultrasound-assisted crystallization (sonocrystallization). Ultrasonics Sonochemistry, 2007. 14: p. 717-724.

[11] Huppertz, T. and I. Gazi, Lactose in dairy ingredients: Effect on processing and storage stability. Journal of Dairy Science, 2015. 99(8): p. 6842-6851.

[12] de Souza, R.R., et al., Recovery and purification of lactose from whey. Chemical Engineering and Processing: Process Intensification, 2010. 49(11): p. 1137-1143.

[13] Bhargava, A., Jelen, P., Lactose Solubility and Crystal Growth as Affected by Mineral Impurities. Journal of Food Science, 1996. 61: p. 180-184.

[14] Siddique, H., et al., Establishment of a Continuous Sonocrystallization Process for Lactose in an Oscillatory Baffled Crystallizer. Organic Process Research & Development, 2015. 19(12): p. 1871-1881.

[15] Kirk, J.H., S.E. Dann, and C.G. Blatchford, Lactose: a definitive guide to polymorph determination. Int J Pharm, 2007. 334(1-2): p. 103-14.

[16] Bund, R. and A. Pandit, Rapid lactose recovery from paneer whey using sonocrystallization: A process optimization. Chemical Engineering and Processing: Process Intensification, 2007. 46: p. 846-850.

[17] Patel, S. and Z. Murthy, Anti-solvent sonocrystallisation of lactose. Chemical and Process Engineering, 2011. 32(4).

Chapter 2

Lactose Synthesis

Lorena Mardones and Marcelo Villagrán

Abstract

This chapter is related to lactose synthesis, its chemistry, regulation, and differences between species, especially in cattle. Lactose synthesis takes place in the Golgi apparatus of mammary epithelial cells (MEC) by the lactose synthase (LS) enzyme complex from two precursors, glucose and UDP-galactose. The enzyme complex is formed by galactosyltransferase, and it is associated with α-lactalbumin. Importantly, the lactose secreted determines the volume of milk produced, due to its osmotic properties. Milk contains 5% lactose and 80% water, percentages that remain constant during lactation in the different mammalian species. The low variation in milk lactose content indicates that lactose synthesis remains constant throughout the period of lactation and that is highly conserved in all mammals. Lactose synthesis is initiated during the first third of the pregnancy, increasing after birth and placenta removal. Different glucose transporters have been involved in mammary glucose uptake, mainly facilitative glucose transporters GLUT1, GLUT8, and GLUT12 and sodium-glucose transporter SGLT1, with more or less participation depending on mammal species.

Keywords: lactose, glucose, glucose transporter, mammary epithelial cells

1. Introduction

The mammary gland plays an essential role during the early postnatal life of young mammals, providing them nutrients, water and electrolytes, and immune protection until they reach the size and maturity to survive independently. The mammary gland has a stroma rich in adipose cells and glandular epithelium that origins a lobule-alveolar system, in which terminal there are alveolar epithelial cells involved in milk production. The mammary gland development, as well as its fundamental structure is very similar among different species, with little differences in function, architecture, and number of glands. For example, in rodents, the branches are few and disperse, whereas ruminants have more branches and they are concentrated in the terminal of alveoli [1]. This gland undergoes cyclic changes that make it reach its maximal development in lactation. This is a unique model of cyclic morphogenesis in adults, with four characteristic steps replicated in each pregnancy and which ends with its involution in menopause. The four phases of mammary gland maturity in adulthood are proliferative phase, secretory differentiation phase, secretory activation phase, and lactation phase [2, 3]. Although breast development begins during embryogenesis, it is during pregnancy when terminal maturation of the gland occurs, developing a lobule-alveolar system characterized by branching of the galactophorous ducts and the differentiation of terminal buds to alveoli. Growth of mammary gland is stimulated during pregnancy by the mammotrophic combination of steroids (estrogen, progesterone, and corticosteroids) and polypeptide hormones (prolactin, growth hormone, and placental lactogen). The four phases of breast differentiation that occur after conception have clear histological differences, as can be seen in **Figure 1**. In the first phase, there is an increase in the amount of glandular tissue, associated with an increase in the number of acini. The second phase is defined by the beginning of lipid synthesis and by its accumulation inside mammary epithelial cells (MEC). The third phase is characterized by the presence of differentiated MEC, capable of producing and secreting all the constituents of milk, which results in a dilation of the alveoli and the presence of an eosinophilic secretion that occupies the acinar lumen. The final phase of mammary gland development is called the lactation phase, and it is the stage in which breastfeeding is established [2]. At this stage, milk secretion is continuous, which is associated with a greater degree of dilation of the alveoli and the presence of milk secretion in the acinar lumen. The process of mammary gland involution starts after weaning and develops in two stages. The first step is a reversible first phase, which lasts a few days, and is due to the release of local breast factors that trigger apoptosis of the secretory epithelia. The second phase, the remodeling, involves the replacing of the lobule-alveolar structures by adipose tissue, degradation of the extracellular matrix and its basal lamina, and the remodeling of adipose tissue [4].

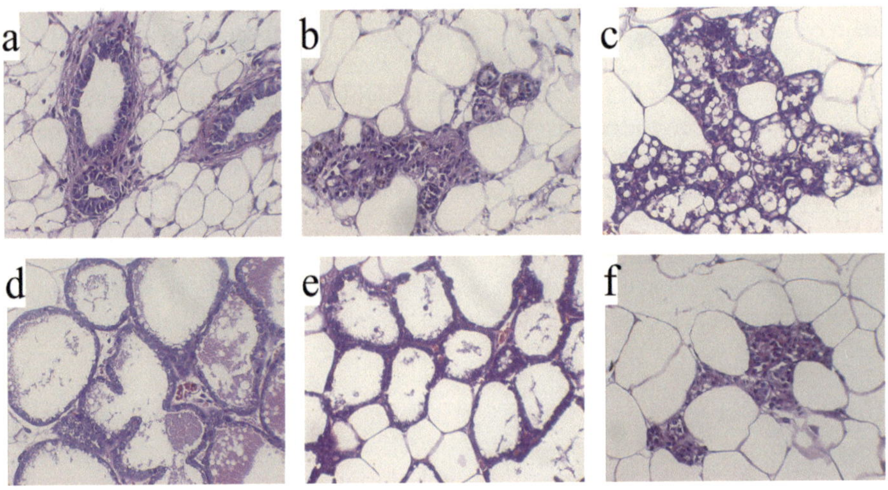

Figure 1.
Histological characteristics of mammary morphogenesis in adults. (a) virgin; (b) proliferative phase; (c) secretory differentiation phase; (d) secretory activation phase; (e) lactation phase; and (f) early involution. Hematoxylin and eosin staining, bar 50 µm [5].

2. Glucose uptake in mammary epithelial cells

Glucose supply to the mammary gland is pivotal to maintain the high rate of proliferation of glandular epithelium in pregnancy and the continuous production of lactose, fat acids, and proteins during lactation [6]. Studies in cows demonstrate that between 60 and 85% of plasmatic glucose is distributed to the mammary gland during lactation and that duodenal glucose injection increases mammary gland glucose uptake and lactose synthesis glucose supply to the mammary gland during lactation, whereas the inhibition of this process or the renal reabsorption of glucose decreased them [7, 8]. On the other hand, in rodents, glucose uptake of the mammary gland duplicates 2 days before delivery, and it remains high during all lactation period, and in humans, 30% of glucose intake is used to lactose production in established lactation [9]. The whole organism adapts to the synthesis of milk; the initial negative energy balance is reversed by greater hepatic gluconeogenesis and decreased peripheral glucose use [10].

The necessity of glucose supply for the proliferation of MEC is very important to lactation persistency, being, together with secretory activity, the two factors that define the lactation curve and peak [4, 11]. In rodents, milk production is mainly associated with a high proliferation of MEC, whereas in cows it is principally due to the increase in secretory activity per cell [12]. In general, MEC proliferation is low in virgin and early pregnancy (5%) in mice, where it has been associated with stem cell renewing, but it persists throughout lactation period, associated with cell replacement, with low net growth, associated with the expression of Ki67 [5, 11]. For example, in cows, the MEC replacement reaches 50%, whereas in rodents it is lower than 25% [11]. When lactation declines, proliferation decreases, and it is exceeded by apoptosis rate, but also the secretory activity by cell decreases [11]. In particular, in humans, mammary growth for lactation starts at week 20 of pregnancy, whereas in mice it starts at day 12 of pregnancy. Studies in mice have established that DNA content increases from the middle of pregnancy until day 5 of lactation, doubling every 6 days, maintaining a net proliferation rate of only 0.3% during lactation, due to parallel apoptosis and loss of cells in milk [13].

The glucose transporters already identified in mammary epithelial cells of rodents, humans, and ruminants are facilitative glucose transporters GLUT1, GLUT8, and GLUT12, SGLT1, and the bidirectional sugar transporter SWEET1 (sugars will eventually be exported transporter) [7, 8, 14–16] (**Figure 2**). The GLUT transporters were initially identified due to the capacity of MEC to transport 3-O-methyl-D-Glucose and inhibition of this by cytochalasin B [17–19]. There are differences in the location and magnitude of peak expression of glucose transporters due to differences between species in the prevalence of cell proliferation or secretory activity [8, 20, 21]. In cows, the increase in the expression of GLUTs is in order of magnitude greater than in rodents, which reflects that its secretory activity

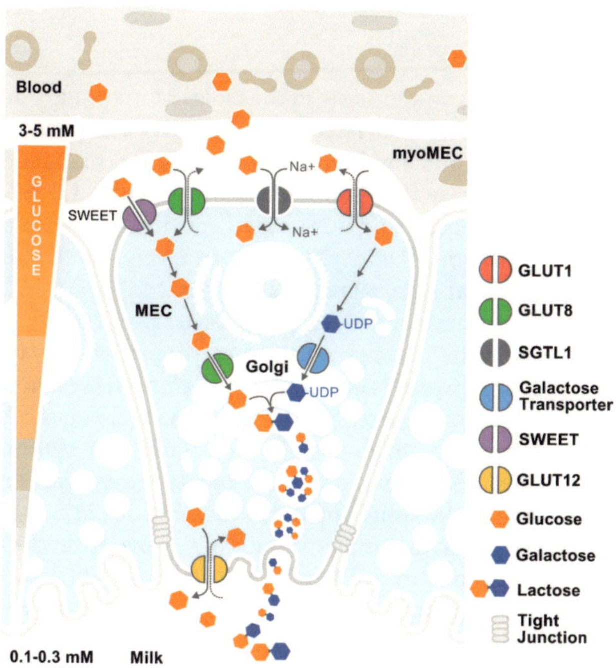

Figure 2.
Glucose uptake in mammary epithelial cell. Distribution of glucose transporters and glucose concentration in different compartments is detailed. MEC, mammary epithelial cells; myoMEC, mammary myoepithelial cells; GLUT, facilitative glucose transporter; SGLT, sodium-glucose cotransporter; SWEET, sugars will eventually be exported transporter.

is highly dependent on the expression of glucose transporters [6, 14]. The maximum expression of GLUT1 observed was between late pregnancy and late lactation, reaching an increase of 5-fold at the protein level and 50-fold at mRNA level [5, 15, 17, 18, 22]. The increase in GLUT8 expression is lower, but it follows the same pattern, associated with cytokeratin 18 and Ki67 expression and MEC proliferation [5, 8]. However, some studies also found an increase of GLUT1 expression in MEC in early pregnancy, which could be related to the start of lipid synthesis in secretory activation phase, when sterol regulatory element-binding protein (SREBP), a transcription factor, appears, or to stem cell renewal [2, 5, 10, 23]. Although the majority of authors found over 60% of GLUT1 expression in plasma membrane in rat lactating gland, some studies found it almost exclusively at intracellular level [5, 19, 20, 24]. Interestingly, in early weaning of BalB/BalC mice, GLUT1 is also concentrated intracellularly, but, due to a decrease in lactation at this step, that could be associated with its accumulation in proteasomal compartment or to apoptotic bodies phagocyted by other epithelial cells acting as nonprofessional phagocytes [5, 25].

The intracellular concentration of glucose in the MEC is mainly determined by its incorporation by GLUT1 transporters in the basolateral membrane and by the activity of the cytosolic hexokinases, which transform glucose into glucose-6-phosphate [2, 23]. The induction of the expression of hexokinase II in the cytoplasm of MEC during the period of breastfeeding is essential in the determination of intracellular glucose levels, because this enzyme has low glucose affinity (K_m 0.3 mM) [26]. On the other hand, glucose is also transported to the lumen of the alveolus, through GLUT12, reaching a concentration of 1.5 mM in milk, equivalent to the concentration found in the cytoplasm of the MEC [20, 26].

3. Lactose synthesis in the Golgi of MEC

3.1 Lactose complex

The first evidence of lactose synthesis in the Golgi of mammary alveolar cells dates back to 1980, when it was associated with the activity of galactosyltransferase and the presence of lactalbumin and bivalent metals such as manganese and calcium [27]. The lactose synthase (LS) synthetizes lactose (beta 1,4-galactoglucose) from UDP-galactose and glucose, and it is located specifically in trans-Golgi. The LS is an enzymatic complex of galactosyltransferase and LALB. LALB is only found in mammary epithelial cells, allowing galactosyltransferase to be specific for the formation of this disaccharide, making galactosyltransferase add galactose to glucose at even low concentration of glucose, increasing its affinity to this carbohydrate 1000-fold [23, 27]. In others cells, galactosyltransferase adds galactose to N-acetylglucosamine glycoconjugates, but in MEC, LALB changes substrate specificity from N-acetylglucosamine to glucose. In fact, the lactose synthesis depends directly on the amount of LALB associated with the galactosyltransferase that is inserted in the inner face of the Golgi apparatus membrane [28]. The LALB knockout produces a viscous, low-lactose milk difficult to remove from the mammary gland, highlighting the osmotic role of lactose in milk yield [29].

LALB expression increases immediately after delivery in pig and rodents and is regulated by lactogenic hormones [30–32]. LS has a K_m of 1.5 mM for glucose and 60 μM for UDP-galactose; thus, the limiting stage in lactose synthesis is the availability of glucose in the Golgi [2, 23, 27]. Lactose synthesis begins in the first third of pregnancy but increases considerably after childbirth, as levels of placental sex steroids decrease, which has an inhibitory effect on lactose synthesis [1]. Lactose production remains relatively constant throughout the entire lactation process thanks to the action of prolactin and other lactogenic hormones and the stimulation associated with mammary gland emptying. The lactose is secreted in the milk together with LALB, α, β, and κ caseins, β-lactoglobulin, others nutrients, and immunomodulatory molecules [1]. Lactose represents around 5% of the milk content in all species, which revealed that is a highly conserved process. Also lactose is an osmotically active molecule, defining the water content in milk, which is in average 80% [1, 7]. In cows, in particular, milk is 5.0% lactose, 3.4% fat, and 2.3% protein. In seals, the lactose content of milk is minimal, and fat is predominant (50%), followed by 6.0% protein. This could be explained because pups need to double their weight in only 4 days to survive adverse environmental conditions [1]. On the other hand, human milk has a similar fat content to cow milk (3.7%), less protein content (1.0%), and more lactose (7.0 v/s. 5.0%). Donkeys presents similar content of macronutrients in milk to human, with 7.4% lactose, 2.0% protein, and 0.4% fat [9].

3.2 Golgi's glucose transporters

The first studies related to glucose transport to the Golgi of MEC concluded that this was mediated by GLUT and SGLT transporters, since the transport of monosaccharides was inhibited by phloretin and phlorizin, known inhibitors of both types of transporters [33]. These vesicles present stereospecificity for several monosaccharides, such as D-glucose, L-glucose, D-xylose, 2-deoxy-D-glucose, and D-fructose. Moreover, the vesicles showed low permeability for glucosamine, a substrate of the GLUT1 transporter; for this reason, we assume that another glucose transporter is involved in the incorporation of glucose into this organelle. A decade later, another GLUT was identified in Golgi vesicles of MEC of late-lactating mice through Western blotting and binding studies of cytochalasin B, co-localizing with 110-kDa coatomer-associated protein β-COP [19, 24]. Interestingly, the results revealed that there is a second cytochalasin B-sensitive glucose transporter, which could correspond to GLUT8, cloned after such studies [34, 35]. In our last study, we were able to identify GLUT8 in the Golgi of lactating MEC in mice, co-localized with LALB, 58 K Golgi protein, and Golgi membrane-associated protein 130 [5]. Additionally, SGLT1 was identified in the Golgi of MEC from lactating cows, but no functional studies have been performed [23]. As SGLT is an active transporter that mobilize glucose thanks to electrochemical sodium gradient at the plasma membrane, more studies related to ion gradient between cytoplasm and inside the Golgi should be performed to really know the contribution of this transporter to glucose uptake into the Golgi of MEC. In **Figure 3**, we show glucose transporters present in the Golgi of mammary epithelial cell and their association with lactose synthesis.

In summary, the reports highlight a variable increase in the expression of GLUT1, GLUT8, and GLUT12 in pregnancy and/or lactation in different models, including rodents and ruminants, but their responsibility in glucose uptake in the Golgi of mammary epithelial cells, an essential step to lactose synthesis, is not clear [8, 24, 36]. Moreover, although GLUT8 has been co-localized with Golgi proteins in MEC and in different compartments of endomembrane system in other cell types, GLUT1 has been found in the Golgi of MEC only in some of the species and strains studied, and its intracellular localization had been associated with mitochondria, which in not part of endomembrane system [5, 24, 33, 37]. In particular, GLUT8 has been found in late endosome and reticulum.

3.3 Lactose synthesis regulation

The lactose synthesis depends principally on lactogenic hormones and glucose uptake in the Golgi of MEC. It starts in the first third of pregnancy, increasing

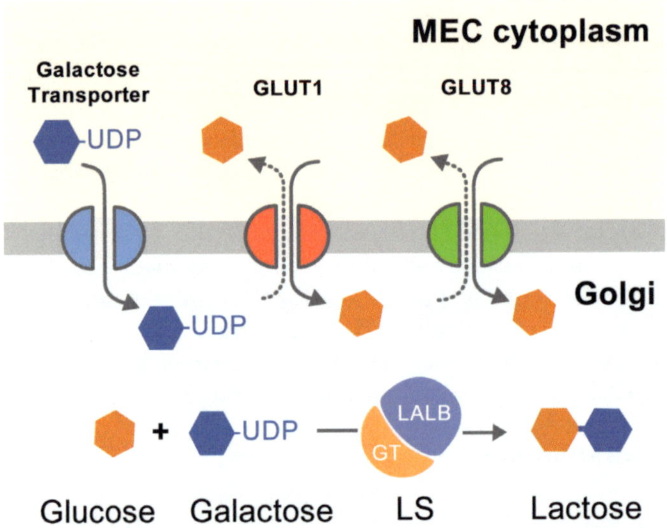

Figure 3.
Glucose transporters in the Golgi of mammary epithelial cell associated with lactose synthesis. Distributions of glucose transporters in this organelle are detailed. MEC, mammary epithelial cells; GLUT, facilitative glucose transporter; LS, lactose synthase; LALB, lactalbumin; GS, galactosyltransferase.

considerately after parturition, when placental sexual steroid hormones decrease and lactogenic hormones increase [30–32]. The principal lactogenic hormone is prolactin, which is stimulated by suckling. The milk production also has been associated with the removal of an inhibitory agent of secretory activity of MEC in milk. Other factors involved in milk production are light and sexual hormones. The increase in light photoperiod from 16 to 18 h increases milk production, due to prolactin via insulin-like growth factor (IGF-1) and somatotropin, whereas pregnancy decreases milk production, due to an increase in estrogens [8]. There is still controversy over the role of LS in milk production. In particular, LALB knockout mice were unable to produce milk, and lactogenic hormones change LALB expression only in particular species. For example, in humans, prolactin increases LALB mRNA, but in rabbits it decreases it, and in other species it does not produce any change [20, 24]. Also, it has been proposed that hexokinase and the different enzymes involved in glucose transformation to UDP-galactose are important for lactose synthesis [7].

As it has been described, the limiting stage in lactose synthesis is the availability of glucose in the Golgi [27], but a combination of lactogenic hormones failed to induce their expression in bovine mammary explants [22, 30]. There are not changes in GLUT8 or GLUT12 expression in response to insulin, leptin, growth hormone, or glucose, but estradiol and progesterone increase GLUT1 in MEC [15, 17, 19, 38]. GLUT1 was redistributed to an intracellular compartment, presumably the Golgi, in response to prolactin and hydrocortisone, associated with phosphatidylinositol-3-kinase (PI3K) and protein kinase C (PKC) pathways and STAT5 binding to its promotor [39–41]. The upregulation of GLUT1 in pregnancy and lactation also has been associated with an increase in serotonin via 5'adenosine monophosphate-activated protein kinase (AMPK) and hypoxia via HIF1α [22, 28]. On the other hand, serotonin increased GLUT8 in the mammary gland and hypoxia and lipopolysaccharide decrease it [10]. GLUT8 is internalized in response to insulin in trophodermic cells and changed its expression in insulin-sensitive tissues, such as the liver and kidney, but failed to produce effects in adipocytes and neuroblast cells [42–44]. Some carbohydrates also produce changes in location and expression of GLUT8, i.e., glucose induces GLUT8 trafficking from the Golgi to the reticulum of hippocampal cells in rats and upregulates it in 3T3-L1 adipocytes [42, 45]. On the other hand, fructose downregulates GLUT8 expression in colon tissue and CaCo-2 colon carcinoma cells but increases its expression in hepatocytes, where it is located in the plasma membrane [46–48]. GLUT8 promoter has a binding sequence to transcription factor NF1, which has been associated with the response of GLUT4 to insulin and cyclic adenosyl monophosphate (cAMP) [42, 49].

4. Conclusion

Mammals rely exclusively on milk supply from the mammary gland to survive at an early age. The proliferation of mammary epithelial cells and mammary establishment depend on glucose supply to the gland, whereas lactose synthesis depends directly on glucose entry into the Golgi of MEC, which is conjugated with UDP-galactose by lactose synthase to produce the disaccharide lactose. MEC presents polarized expression of GLUT1, SGLT1, and GLUT12 in its plasma membrane and also expresses GLUT1, GLUT8, and SGLT1 in the Golgi. Hormones and oxygen tension regulate the expression of these transporters; however, further studies are necessary to explore the effects of light/dark cycles and suckling in their expression, since these are factors involved in milk production. Additionally, the kinetics of transporters involved in glucose uptake in the Golgi or cytoplasm of MEC also needs to be explored. Understanding the regulation and function of glucose transporters will be useful to improve efficiency of milk yield in both, humans and cattle.

Acknowledgements

The authors would like to thank Fredy Díaz of Interactive Biology Laboratory, Faculty of Medicine at Universidad Católica de la Santísima Concepción, Concepción, Chile, for drawing the figures we designed and the Vice-Rectory for Research and Post-grade at Universidad Católica de la Santísima Concepción for partial finance of this publication.

Conflict of interest

The authors declare no conflict of interest.

Nomenclature

MEC	mammary epithelial cell
GLUT	facilitative glucose transporter
SGLT	sodium-glucose transporter
SWEET	sugars will eventually be exported transporter
LALB	lactalbumin
LS	lactose synthase
SREBP	sterol regulatory element-binding protein
K_m	Michaelis-Menten kinetic constant
β-COP	110-kDa coatomer-associated protein
AMPK	5' adenosine monophosphate-activated protein kinase

PI3K	phosphatidylinositol-3-kinase
cAMP	cyclic adenosyl monophosphate
NF1	nuclear factor 1
STAT5	signal transducer and activator of transcription 5
PKC	protein kinase C

Author details

Lorena Mardones* and Marcelo Villagrán
Unit of Nutritional Biochemistry, Biomedical Sciences Research Laboratory, Faculty of Medicine, Universidad Católica de la Santísima Concepción, Concepción, Chile

*Address all correspondence to: lmardones@ucsc.cl

References

[1] Akers R. Lactation and the Mammary Gland. 1st ed. Iowa: Blackwell Publishing Company; 2002. 278 p. DOI: 10.1002/9781119264880

[2] Anderson SM, Rudolph MC, McManaman JL, Neville MC. Key stages in mammary gland development. Secretory activation in the mammary gland: It's not just about milk protein synthesis! Breast Cancer Research. 2007;**9**(1):204. DOI: 10.1186/bcr1653

[3] Russo J, Russo IH. Development of the human breast. Maturitas. 2004;**49**(1):2-15. DOI: 10.1016/j.maturitas.2004.04.011

[4] Davies CR, Morris JS, Griffiths MR, Page MJ, Pitt A, Stein T, et al. Proteomic analysis of the mouse mammary gland is a powerful tool to identify novel proteins that are differentially expressed during mammary development. Proteomics. 2006;**6**(21):5694-5704. DOI: 10.1002/pmic.200600202

[5] Villagran M, Munoz M, Inostroza E, Venegas C, Ruminot I, Parra-Valencia E, et al. GLUT1 and GLUT8 support lactose synthesis in Golgi of murine mammary epithelial cells. Journal of Physiology and Biochemistry. 2019;**75**(2):209-215. DOI: 10.1007/s13105-019-00679-3

[6] Capuco AV, Ellis SE, Hale SA, Long E, Erdman RA, Zhao X, et al. Lactation persistency: Insights from mammary cell proliferation studies. Journal of Animal Science. 2003;**81**(Suppl 3):18-31

[7] Xiao CT, Cant JP. Relationship between glucose transport and metabolism in isolated bovine mammary epithelial cells. Journal of Dairy Science. 2005;**88**(8):2794-2805. DOI: 10.3168/jds.S0022-0302(05)72959-3

[8] Zhao FQ. Biology of glucose transport in the mammary gland. Journal of Mammary Gland Biology and Neoplasia. 2014;**19**(1):3-17. DOI: 10.1007/s10911-013-9310-8

[9] Lawrence R, Lawrence R. Breastfeeding: A Guide for the Medical Profession. 8th ed. Missouri: Elsevier Inc.; 2015, 992 pp. DOI: 10.1111/apa.13474

[10] Laporta J, Peters TL, Merriman KE, Vezina CM, Hernandez LL. Serotonin (5-HT) affects expression of liver metabolic enzymes and mammary gland glucose transporters during the transition from pregnancy to lactation. PLoS One. 2013;**8**(2):e57847. DOI: 10.1371/journal.pone.0057847

[11] Capuco AV, Ellis SE. Comparative aspects of mammary gland development and homeostasis. Annual Review of Animal Biosciences. 2013;**1**:179-202. DOI: 10.1146/annurev-animal-031412-103632

[12] Richmond PA, Hood LF. Tissue culture method for bovine mammary tissue: Lactose synthesis. Journal of Dairy Science. 1973;**56**(5):611-614. DOI: 10.3168/jds.S0022-0302(73)85228-2

[13] Knight CH, Peaker M. Mammary cell proliferation in mice during pregnancy and lactation in relation to milk yield. Quarterly Journal of Experimental Physiology. 1982;**67**(1):165-177

[14] Zhao FQ, Keating AF. Functional properties and genomics of glucose transporters. Current Genomics. 2007;**8**(2):113-128

[15] Zhao FQ, Keating AF. Expression and regulation of glucose transporters in the bovine mammary gland. Journal of Dairy Science. 2007;**90**(Suppl 1):E76-E86. DOI: 10.3168/jds.2006-470

[16] Zhu LQ, Bao ZK, Hu WW, Lin J, Yang Q, Yu QH. Cloning and functional analysis of goat SWEET1. Genetics and Molecular Research. 2015;**14**(4):17124-17133. DOI: 10.4238/2015.December.16.12

[17] Burnol AF, Leturque A, Loizeau M, Postic C, Girard J. Glucose transporter expression in rat mammary gland. The Biochemical Journal. 1990;**270**(1):277-279

[18] Prosser CG. Mechanism of the decrease in hexose transport by mouse mammary epithelial cells caused by fasting. The Biochemical Journal. 1988;**249**(1):149-154

[19] Madon RJ, Martin S, Davies A, Fawcett HA, Flint DJ, Baldwin SA. Identification and characterization of glucose transport proteins in plasma membrane- and Golgi vesicle-enriched fractions prepared from lactating rat mammary gland. The Biochemical Journal. 1990;**272**(1):99-105. DOI: 10.1042/bj2720099

[20] Macheda ML, Williams ED, Best JD, Wlodek ME, Rogers S. Expression and localisation of GLUT1 and GLUT12 glucose transporters in the pregnant and lactating rat mammary gland. Cell and Tissue Research. 2003;**311**(1):91-97. DOI: 10.1007/s00441-002-0661-5

[21] Yu Q, Zhu L, Lin J, Zhang Q, Tian Q, Hu W, et al. Functional analyse of GLUT1 and GLUT12 in glucose uptake in goat mammary gland epithelial cells. PLoS One. 2013;**8**(5):e65013. DOI: 10.1371/journal.pone.0065013

[22] Komatsu T, Itoh F, Kushibiki S, Hodate K. Changes in gene expression of glucose transporters in lactating and nonlactating cows. Journal of Animal Science. 2005;**83**(3):557-564

[23] Zhao FQ, Okine EK, Kennelly JJ. Glucose transporter gene expression in bovine mammary gland. Journal of Animal Science. 1999;**77**(9):2517-2522

[24] Nemeth BA, Tsang SW, Geske RS, Haney PM. Golgi targeting of the GLUT1 glucose transporter in lactating mouse mammary gland. Pediatric Research. 2000;**47**(4 Pt 1):444-450

[25] Atabai K, Sheppard D, Werb Z. Roles of the innate immune system in mammary gland remodeling during involution. Journal of Mammary Gland Biology and Neoplasia. 2007;**12**(1):37-45. DOI: 10.1007/s10911-007-9036-6

[26] Anderson ME. Determination of glutathione and glutathione disulfide in biological samples. Methods in Enzymology. 1985;**113**:548-555

[27] Kuhn NJ, Wooding FB, White A. Properties of galactosyltransferase-enriched vesicles of Golgi membranes from lactating-rat mammary gland. European Journal of Biochemistry. 1980;**103**(2):377-385

[28] Akers RM, Bauman DE, Capuco AV, Goodman GT, Tucker HA. Prolactin regulation of milk secretion and biochemical differentiation of mammary epithelial cells in periparturient cows. Endocrinology. 1981;**109**(1):23-30. DOI: 10.1210/endo-109-1-23

[29] Stacey A, Schnieke A, Kerr M, Scott A, McKee C, Cottingham I, et al. Lactation is disrupted by alpha-lactalbumin deficiency and can be restored by human alpha-lactalbumin gene replacement in mice. Proceedings of the National Academy of Sciences of the United States of America. 1995;**92**(7):2835-2839. DOI: 10.1073/pnas.92.7.2835

[30] Shao Y, Wall EH, McFadden TB, Misra Y, Qian X, Blauwiekel R, et al. Lactogenic hormones stimulate expression of lipogenic genes but

not glucose transporters in bovine mammary gland. Domestic Animal Endocrinology. 2013;**44**(2):57-69. DOI: 10.1016/j.domaniend.2012.09.001

[31] Molenaar AJ, Davis SR, Wilkins RJ. Expression of alpha-lactalbumin, alpha-S1-casein, and lactoferrin genes is heterogeneous in sheep and cattle mammary tissue. The Journal of Histochemistry and Cytochemistry. 1992;**40**(5):611-618

[32] Burditt LJ, Parker D, Craig RK, Getova T, Campbell PN. Differential expression of alpha-lactalbumin and casein genes during the onset of lactation in the Guinea-pig mammary gland. The Biochemical Journal. 1981;**194**(3):999-1006

[33] White MD, Kuhn NJ, Ward S. Permeability of lactating-rat mammary gland Golgi membranes to monosaccharides. The Biochemical Journal. 1980;**190**(3):621-624

[34] Doege H, Schürmann A, Bahrenberg G, Brauers A, Joost HG. GLUT8, a novel member of the sugar transport facilitator family with glucose transport activity. The Journal of Biological Chemistry. 2000;**275**(21):16275-16280

[35] Ibberson M, Uldry M, Thorens B. GLUTX1, a novel mammalian glucose transporter expressed in the central nervous system and insulin-sensitive tissues. The Journal of Biological Chemistry. 2000;**275**(7):4607-4612

[36] Miller PJ, Finucane KA, Hughes M, Zhao FQ. Cloning and expression of bovine glucose transporter GLUT12. Mammalian Genome. 2005;**16**(11):873-883. DOI: 10.1007/s00335-005-0080-5

[37] Diril MK, Schmidt S, Krauss M, Gawlik V, Joost HG, Schürmann A, et al. Lysosomal localization of GLUT8 in the testis—The EXXXLL motif of GLUT8 is sufficient for its intracellular sorting via AP1- and AP2-mediated interaction. The FEBS Journal. 2009;**276**(14):3729-3743. DOI: 10.1111/j.1742-4658.2009.07089.x

[38] Zhao FQ, Dixon WT, Kennelly JJ. Localization and gene expression of glucose transporters in bovine mammary gland. Comparative Biochemistry and Physiology. Part B, Biochemistry & Molecular Biology. 1996;**115**(1):127-134

[39] Riskin A, Nannegari VH, Mond Y. Acute effectors of GLUT1 glucose transporter subcellular targeting in CIT3 mouse mammary epithelial cells. Pediatric Research. 2008;**63**(1):56-61. DOI: 10.1203/PDR.0b013e31815b440b

[40] Zhao FQ, Moseley WM, Tucker HA, Kennelly JJ. Regulation of glucose transporter gene expression in mammary gland, muscle, and fat of lactating cows by administration of bovine growth hormone and bovine growth hormone-releasing factor. Journal of Animal Science. 1996;**74**(1):183-189

[41] Zhao K, Liu HY, Zhao FQ, Liu JX. Short communication: Protein kinase C regulates glucose uptake and mRNA expression of glucose transporter (GLUT) 1 and GLUT8 in lactating bovine mammary epithelial cells. Journal of Dairy Science. 2014;**97**(7):4329-4332. DOI: 10.3168/jds.2013-7848

[42] Scheepers A, Doege H, Joost HG, Schürmann A. Mouse GLUT8: Genomic organization and regulation of expression in 3T3-L1 adipocytes by glucose. Biochemical and Biophysical Research Communications. 2001;**288**(4):969-974. DOI: 10.1006/bbrc.2001.5866

[43] Schmidt S, Joost HG, Schürmann A. GLUT8, the enigmatic intracellular hexose transporter. American Journal

of Physiology. Endocrinology and Metabolism. 2009;**296**(4):E614-E618. DOI: 10.1152/ajpendo.91019.2008

[44] Schürmann A, Joost HG, Cushman SW, Al-Hasani H. Targeting of GLUT6 (formerly GLUT9) and GLUT8 in rat adipose cells. The Biochemical Journal. 2001;**358**(Pt 2):517-522

[45] Piroli GG, Grillo CA, Hoskin EK, Znamensky V, Katz EB, Milner TA, et al. Peripheral glucose administration stimulates the translocation of GLUT8 glucose transporter to the endoplasmic reticulum in the rat hippocampus. The Journal of Comparative Neurology. 2002;**452**(2):103-114. DOI: 10.1002/cne.10368

[46] DeBosch BJ, Chen Z, Finck BN, Chi M, Moley KH. Glucose transporter-8 (GLUT8) mediates glucose intolerance and dyslipidemia in high-fructose diet-fed male mice. Molecular Endocrinology. 2013;**27**(11):1887-1896. DOI: 10.1210/me.2013-1137

[47] Debosch BJ, Chen Z, Saben JL, Finck BN, Moley KH. Glucose transporter 8 (GLUT8) mediates fructose-induced de novo lipogenesis and macrosteatosis. The Journal of Biological Chemistry. 2014;**289**(16):10989-10998. DOI: 10.1074/jbc.M113.527002

[48] DeBosch BJ, Chi M, Moley KH. Glucose transporter 8 (GLUT8) regulates enterocyte fructose transport and global mammalian fructose utilization. Endocrinology. 2012;**153**(9):4181-4191. DOI: 10.1210/en.2012-1541

[49] Miura S, Tsunoda N, Ikeda S, Kai Y, Cooke DW, Lane MD, et al. Nuclear factor 1 regulates adipose tissue-specific expression in the mouse GLUT4 gene. Biochemical and Biophysical Research Communications. 2004;**325**(3):812-818. DOI: 10.1016/j.bbrc.2004.10.117

Section 2

Cheese Whey as a Source of Lactose and Lactose Derivatives

Chapter 3

Membrane Applications for Lactose Recovering

Antónia Teresa Zorro Nobre Macedo,
Joana Filipa Oliveira Monteiro,
David José Chaveiro da Silva Azedo,
Elizabeth da Costa Neves Fernandes de Almeida Duarte
and Carlos Dias Pereira

Abstract

Cheese whey, the co-product from cheese making processes, is a natural and cheap source of high value compounds, mainly proteins, small peptides, oligosaccharides, lactose, and minerals. Lactose is the main component (about 90%) of the dry extract of cheese whey. This carbohydrate has plenty of application in the food and pharmaceutical industries due to its relative low sweetening power, caloric value, and glycemic index. Besides, lactose is currently available for diverse physicochemical properties, namely particle size, bulk density, distribution, and flow characteristics, extending its use for a larger range of applications. Recovery of lactose from cheese whey can be carried out through different processes, such as membrane processes, crystallization, anti-solvent crystallization, and sonocrystallization. This chapter aims to furnish a deep insight into the performance of membrane processes for lactose recovery from cheese whey.

Keywords: cheese whey, lactose recovery, membrane processes, nanofiltration, ultrafiltration

1. Introduction

Dairy industry is one of the major food processing industries in the world, manufacturing a broad range of different products. Therefore, it generates large amounts of by-products during the processing of milk and manufacture of dairy products (e.g., cheese, butter, and yogurts), leading to problems of their management/utilization [1].

Cheese whey is the most abundant co-product in the cheese-making and casein industries. It contains about 65 g L^{-1} of dry matter, being lactose the main component (70–80%), proteins (9%), corresponding to 20% of all milk proteins, and minerals (8–20%) and, to a much lesser extent, hydrolyzed peptides from casein-k, lipids, and bacteria, which resulted from cheese manufacturing [2, 3]. Generally, for each 100 kg of milk, around 10–20 kg of cheese is manufactured, and 80–90 kg of liquid whey is released [4]. According to Food and Agriculture Organization Corporate Statistical Database (FAOSTAT), more than 114 million tons of whey were produced worldwide in 2013, with Europe producing 63 million tons in that year [5]. Data from the European Whey Products Association (EWPA) indicated that about 6 million tons of whey (dry matter) were produced in the European Union in the year 2015 [6]. In spite of these larger volumes produced, only around 50% of the whey annually produced in the world is valorized into different added-value products. This is because, although cheese whey is an inexpensive and abundant source for developing new added-value products (e.g., foods, pharmaceuticals, and energy), its low solid content makes it difficult for direct utilization [4]. Therefore, for recovering any of its components, such as the lactose, several processes, mainly separation processes, should be used. The intended final use of lactose determines the process that should be used for its separation from cheese whey.

2. Membrane processes

Membrane separation is a filtration process based on the use of membranes for the separation of dissolved or colloidal solids in liquid mixtures, or the separation of small components in gaseous mixtures. A membrane is a permselective barrier between two phases (feed/retentate) and permeate, which preferably allows the permeation of a component (or components) of the feed retaining others, leading to their separation, purification, or concentration. The difference in permeability (membrane transport) between the components of the mixture is due to differences in size (ratio between mean pore radius of membrane and size of solute to be separated) and/or chemical selectivity for membrane material (relationship among chemical characteristics) [4].

These processes differ from frontal filtration in the following characteristics: (1) the particle size they separate; (2) tangential rather than dead-end mode of feed introduction; and (3) use of membranes, in spite of depth filters. Therefore, these processes allow to expand the scope of frontal filtration for separating components of smaller dimensions (less than 1 μm). The parallel flow limits the accumulation of substances retained on the membrane due to shear stress and two different product streams are obtained (**Figure 1**). When using membranes, the components are retained to the surface in a thin film, called the active layer or skin, and so higher retention rates are possible [4, 7].

Membrane separation processes can be classified according to the driving force that controls the mass transfer rate of the individual components from one phase to another. These driving forces can be of several natures such as concentration gradients, temperature, pressure, and external force fields. The main processes used at an industrial level are pressure-driven processes, such as microfiltration, ultrafiltration, nanofiltration, and reverse osmosis [8, 9]. In these processes, by applying a pressure, the solvent and some solutes freely permeate the membrane, while others are retained to varying extents, depending on various factors, such as solute, membrane characteristics, operating parameters, or others [8, 9]. The size of the particle or molecule to be separated as well as its chemical properties determines the

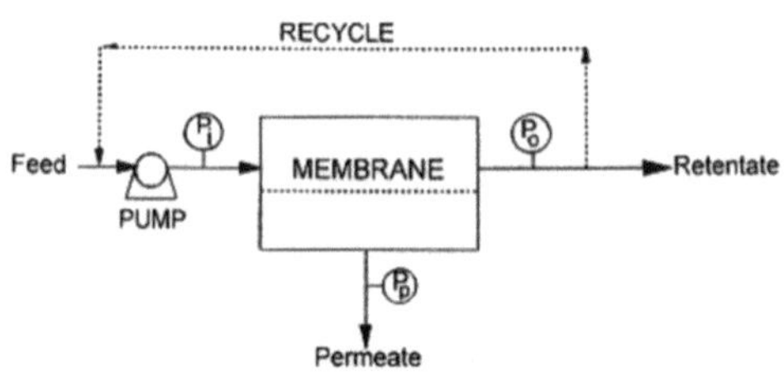

Figure 1.
Diagram of a membrane separation process [4].

structure (porous or dense, pore size, and pore size distribution) of the membrane to be used. The nature of the solvent (aqueous or organic), the cleaning method, the applied pressure, and the temperature influence the type of membrane material [10]. When it progresses toward microfiltration, ultrafiltration, nanofiltration, and reverse osmosis, the size or molecular weight of particles or molecules that are retained by the membrane, pore size, and porosity decreases. This means that the hydrodynamic resistance of the membranes to mass transfer is increasing, requiring higher applied pressures to achieve the same permeation fluxes.

Following water and wastewater treatment, the food industry ranks second in applications of those processes. Most applications are in the dairy industry (production of whey protein concentrates; milk protein standardization), followed by the beverage (wine, beer, vinegar, and fruit juice) and egg product industries [8, 11]. In the food industry, the application of membrane separation processes provides several benefits, such as food safety, competitiveness, innovation, and environmental compatibility. Food safety through membrane processes can be achieved, for example, by cold sterilization, using microfiltration. They are competitive with other concentration processes, for example, thermal processes, due to their lower energy consumption. In addition, they can be easily integrated into industrial plants due to ease of implementation, possibility of using compact modules, and good automation. So, these processes are currently present in several industrial plants, namely in the development of new value-added products, for example, from by-products (cheese whey or second cheese whey) and/or residues of the food industry. In addition, since only cleaning agents are used and the processes can be operated under mild conditions (pressure and temperature), they are recognized as green processes [3].

2.1 Membranes

The membranes can be manufactured with different types of materials (polymeric or inorganic), may have different structures (symmetrical or asymmetrical), and are usually commercialized in arrangements of membranes, with a high surface area per unit volume, called modules.

The nature of the material used is an important aspect of membrane processes because it can affect the behavior and performance and limits the use of a membrane, for a particular application. Regardless of its nature, that material must have good thermal, mechanical, and chemical stability; hydrophilicity or hydrophobicity; ease of manufacture on a wide variety of dimensions pores; modules; and configurations [4, 7]. In this respect, inorganic membranes made from ceramic materials are the most used, due to its higher thermal, chemical, and mechanical stability than polymeric membranes. These characteristics allow its use in a wider pH region and with different organic solvents. Furthermore, they are easier to clean and disinfect, since more concentrated solutions of strong acids and bases and higher temperatures can be used, keeping their life span. Some disadvantages of these membranes compared to polymeric ones are mainly associated with its higher cost, the need of using higher flow rates (greater energy consumption), and to the fact that, currently, does not exist in the market ceramic nanofiltration membranes with limit of separation less than 250 Da [12].

The classification of membranes according to their structure is shown schematically in **Figure 2**. Symmetrical membranes include microporous and homogeneous membranes (dense and nonporous). The thickness of the symmetric membranes can vary approximately from 10 to 200 μm, the resistance to mass transfer being determined by the total thickness of the membrane. Thus, the thinner the membrane, the higher the permeation rate [7]. These membranes are applied in microfiltration and can be classified, on an absolute scale, through their maximum

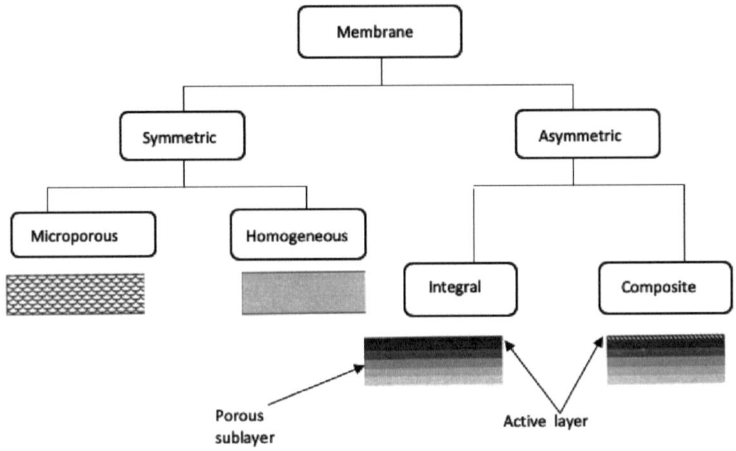

Figure 2.
Schematic representation of membrane structure.

equivalent pore diameter. Homogeneous membranes are mainly applied in gas separation. Asymmetric membranes have a structure consisted of a very thin film on their surface with a thickness in the range of 0.1–0.5 μm, called skin or active layer, which is based on a porous support layer, the thickness of which can vary between 100 and 200 μm [4, 7]. The separation occurs only at the surface, in the active layer, retaining components whose molar mass is greater than the molecular weight cut-off (MWCO) of the membrane, which is defined as the molar mass that is 90% rejected by this membrane. The manufacturing process of the membranes still leads to obtaining two different substructures: the integral asymmetric membrane design and nonintegral asymmetric membranes, the latter forming part of the composite membranes. Integral asymmetric membranes are obtained from a single polymer. Composite membranes, also called thin-film, thin, or ultrafine layer composites, are manufactured with a polymer (or other material) different from that used in the layer support and in several stages, which make it possible to optimize each of them, independently. These membranes are used in ultrafiltration, nanofiltration, and reverse osmosis.

The design of the modules is based on two types of membrane configurations: flat and tubular. Plate modules and spiral-wound modules involve flat membranes, while tubular, capillary, and hollow fiber modules are based on tubular membrane configurations. In general, an industrial membrane installation consists of the association of several modules, which are selected and configured in parallel or in series, depending on the production/specification of the final product. The selection of the module configuration, as well as the module arrangement, is based on several factors: economic considerations; type of application; ease of cleaning, maintenance, and operation; compactness of the system; and scale and possibility of replacing membranes.

2.2 Performance of membrane processes

The main parameters used to evaluate the performance of a membrane are the permeate flux that is a measure of its productivity and the apparent rejection coefficient, which allows us to estimate their selectivity. The permeate flux (J_v) is defined as the amount, in volume or mass, that passes through the membrane per unit area and time, that is,

$$J_v = \frac{V}{A \times t} \left(\mathrm{m\ s^{-1}}\right) \quad (1)$$

where J_v is the volumetric permeate flux ($\mathrm{m\ s^{-1}}$); V is the volume of the permeate ($\mathrm{m^3}$); A is the membrane surface ($\mathrm{m^2}$); and t (s) is the time required to collect the volume of permeate V.

The rejection coefficient is a measure of membrane selectivity for the separation of a given solute, which may be partially or totally retained by it, while the solvent freely permeates the membrane. The apparent (or observed) rejection coefficient, R, is defined as follows:

$$R = \frac{C_f - C_p}{C_f} \quad (2)$$

where C_f is the concentration of a particular solute in the feed, and C_p is the concentration of this solute in the permeate.

The apparent rejection coefficient depends on the experimental conditions, namely transmembrane pressure and feed circulation velocity. This coefficient is a dimensionless quantity, which can take values between 0 and 1, as the solute freely permeates the membrane or is completely retained by it, respectively. The latter situation corresponds to an ideal semi-permeable membrane.

The permeate flux and apparent rejection coefficient are influenced by several factors related to solute characteristics (size and shape, macro and micro solute coexistence), membranes (more hydrophobic/hydrophilic character, surface charge distribution, and surface roughness), operating parameters (transmembrane pressure, feed circulation velocity, and temperature), environmental conditions (pH, ionic strength, and osmotic pressure), and type of module (plane and tubular) [13]. These factors give rise to the resistive phenomena mass transfer across the membrane, referred to as concentration polarization and fouling, which can severely affect the performance of membrane processes. Another factor that can reduce the performance of membrane processes is the effect of osmotic pressure.

Concentration polarization consists in the formation of a concentration gradient in a thin layer near the membrane surface, caused by the accumulation of the retained species and leads to the initial decrease of permeate fluxes, which may also contribute to a reduced selectivity. It mainly affects those processes with larger pore membranes (higher permeate fluxes) such as microfiltration and ultrafiltration and can be minimized through the use of low pressures, high feed circulation rates, and low solute concentrations.

Fouling consists of pore obstruction (on or in the surface), caused by solute-membrane or solute-solute interactions, which mainly depend on the characteristics of the solutes, of the membrane, and of operating conditions and the type of module. This phenomenon can lead to a sharp reduction in permeate flux and can alter membrane selectivity [14]. In order to reduce the effects of fouling, various preventive methods can be used such as (1) use a suitable pre-treatment for the food (pre-filtration, pH adjustment, and adequate heat treatment); (2) select the most suitable membrane (narrow pore size distribution, hydrophobicity characteristics, presence of charged groups, or with certain functional characteristics on the membrane surface); (3) use the modules with spacers and work with high feed circulation rates or even at low permeate fluxes, by reducing the applied transmembrane pressure; and (4) use the rotary (or vibratory) modules, in which the membrane moves on a rotating cylinder, creating greater turbulence close to the membrane, compared with conventional tangential modules, while maintaining low shear rates within the fluid [15].

The effect of osmotic pressure in the decline of permeate fluxes is generally neglected in microfiltration and ultrafiltration since the solutes to be separated in these cases have very high molar masses. However, if the concentration of macromolecular solutes is very high, then this effect will have to be accounted for. The phenomenon is especially important in reverse osmosis and also nanofiltration, since in these processes, the solutes that separate are of low molar mass, so the osmotic pressures can be high, decreasing the effective pressure.

In addition, the performance of the overall membrane process should also take into account economic factors, such as membrane prices and shelf life, cleaning and disinfection reagents, and energy consumption.

3. Lactose recovery through membrane processes

In the industrial process that is currently used for lactose production, membrane separation techniques have already been introduced because lactose is currently recovered from the whey ultrafiltration permeate. The whey proteins separated have different and interesting applications (e.g., whey protein concentrates, WPC, or whey protein isolates, WPI), thus contributing to the valorization of cheese whey. The Commonwealth Scientific and Industrial Research Organization (CSIRO) developed a method for the possible commercial production of pharmaceutical quality lactose, which integrates the following operations: ion exchange (for calcium and magnesium removal), nanofiltration/diafiltration (for lactose separation, concentration, and purification), evaporation, crystallization, and chromatography, allowing not only to obtain high purity lactose, as well as mineral salts and calcium from cheese whey. This process has several benefits because through the use of nanofiltration/diafiltration, it is produced by a purified lactose concentrate, minimizing simultaneously the evaporation costs due to the reducing volume. Besides, the nanofiltration permeate can be subjected to reverse osmosis, producing water of good quality (e.g., for cleaning and diafiltration).

The recovery of lactose from cheese whey by membrane processes is mainly carried out by nanofiltration (NF) of the ultrafiltration permeates, due to their physical-chemical composition. Those permeates are composed of small solutes, being lactose the major compound of the dry matter, followed by several ions such as, sodium, potassium, calcium, magnesium, chloride, phosphate, and citrate. Therefore, the specific selectivity of NF to this type of solutes and its lower energy consumption, compared with other processes such as reverse osmosis and evaporation, has boosted their use in dairy [8, 16, 17] and other agroindustrial sectors.

One of the most important uses of nanofiltration is the production of whey-demineralized lactose concentrates in the food industry, or even, if enough purification is achieved, for pharmaceutical purposes. The demineralization of dairy fluids is very important to reduce their high salt content (8–20% of dry matter) [3, 18], which causes several difficulties in processing. A high salt content leads to slow lactose crystallization rate because it reduces lactose solubility in supernatant liquor during crystallization.

The major drawback of the NF process is the fouling caused by mineral precipitation of salts, namely calcium phosphates. Another reason for the decrease of permeate flux is the increase of osmotic pressure and concentration polarization, due to the accumulation of lactose and salts (sodium, potassium, and chloride) near the membrane surface, causing a reduction in the effective pressure [19, 20].

Guu and co-workers [21] found that the application of NF for sweet whey or UF permeates allowed to increase the production of lactose crystals by about 10 and 8%, respectively, for a VRR of 3.0. This behavior was attributed to the partial demineralization of the permeate, especially in terms of the monovalent ions, sodium and potassium. These results raised the interest for the integration of NF membranes in the lactose production plants at the industrial level.

Rice and co-workers [22] carried out nanofiltration of ultrafiltration permeates using polyamide membranes NF270 and observed a severe flux decline during filtration at high temperatures and pH, due to calcium phosphate precipitation, because of its lower solubility in these operating conditions. Those authors suggested that if the pH of the feed was reduced, fouling could be avoided, despite changing the separation properties of the membrane.

Cuartas-Uribe and co-workers [23] studied the concentration of lactose from whey ultrafiltration permeates, combining concentration by nanofiltration with continuous diafiltration modes, and found that the best operating conditions were a transmembrane pressure of 2.0 MPa and a volume dilution factor of around 2.0 because a good removal of chloride was possible with the lowest lactose loss for the permeate. Although these authors claimed that no fouling problems were detected during NF tests, experiments at a larger scale should be performed to evaluate the economic feasibility of the process.

Ferg and co-workers [24] also investigated the recovery of lactose through a combination of membrane processes, namely MF (nominal pore size 0.2 μm), UF (5 kDa MWCO), ion exchange, and RO, and obtained an overall lactose recovery of 74%, with a purity of 99.8%.

Bertoluzzi and co-workers [25] compared the performance of two double-stage membrane processes for treatment of dairy wastewaters: (1) microfiltration (MF) plus NF and (2) MF plus OI. For MF, a hollow fibber module was used, being membranes made of poly(ether sulfonate)/poly(vinyl pyrrolidone) (PES/PVP) mixture with a pore size of 0.20 μm. In the NF and RO experiments, polymeric flat-type membranes were used, being these membranes made of polyamide composites. For the NF experiments, they used two different membranes (NF90 and NF), which are made of the same material but have different rejection properties, since NF90 is a tighter membrane, while the other one is a looser membrane, as can also be confirmed by their hydraulic permeabilities to pure water. Before the experiments, the dairy wastewater was prefiltrated across a filter of 0.25 μm to remove solids and to avoid a quick fouling of membranes. After that, microfiltration was also used as a pretreatment for the next operation (NF or OI) with the objective of improving their performance. The authors found that the sequence of MF followed by RO allowed a better removal of total solids and organic matter. Besides, the composition of the final permeate was compatible with the discharge on receiving waters according to the Brazilian environmental regulations or could be used in cleaning-in-place processes in the dairy factory. Although the results of this study are a good basis for other similar dairy wastewaters, since the variety of manufacturing processes involved in dairy products used is too large, for each type of sample/desired goal, a previous study is always necessary.

Macedo and co-workers [20] used a combination of UF/NF and UF/DF followed by NF/DF of the previous permeates, to recover lactose from the permeates both of sheep cheese whey (PUF-S) and of goat cheese whey (PUF-G) (**Figure 3**).

Both samples were subjected to the following pretreatment: filtration (using traditional cotton cloths), skimming for fat removal, and low pasteurization. NF of both permeates was carried out with NFT50 (NF) membranes until a volume concentration factor (VCF) of about 2.5. It was observed a sharp decrease (around 60%) in the permeate flux in the case of PUF-S and a smaller reduction (about 20%) in PUF-G (**Figure 4**). The authors attributed this different behavior to the following factors: the higher concentration of lactose and applied pressure used in the case of PUF-S (higher permeate fluxes) led to a greater and faster accumulation

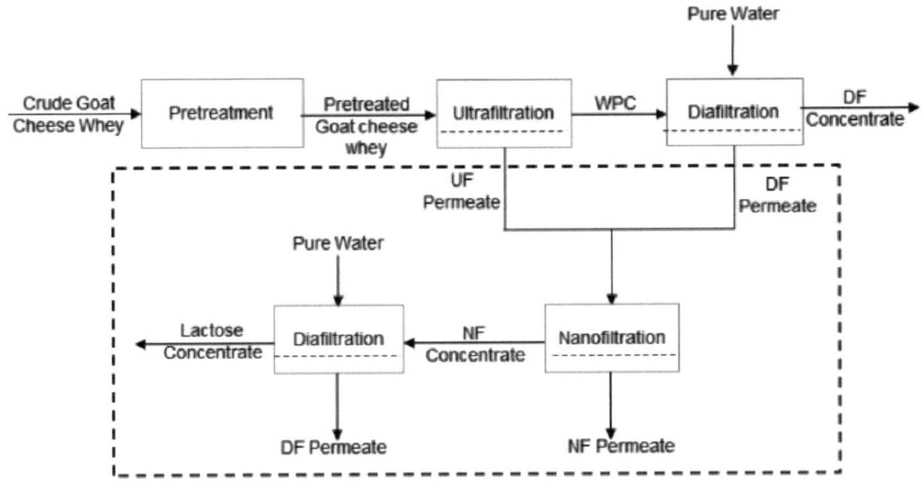

Figure 3.
Recover of lactose (lactose concentrate) and whey proteins from cheese whey: WPC = whey protein concentrate; DF concentrate = whey protein concentrate of UF/DF; lactose concentrate (obtained after NF/DF) [20].

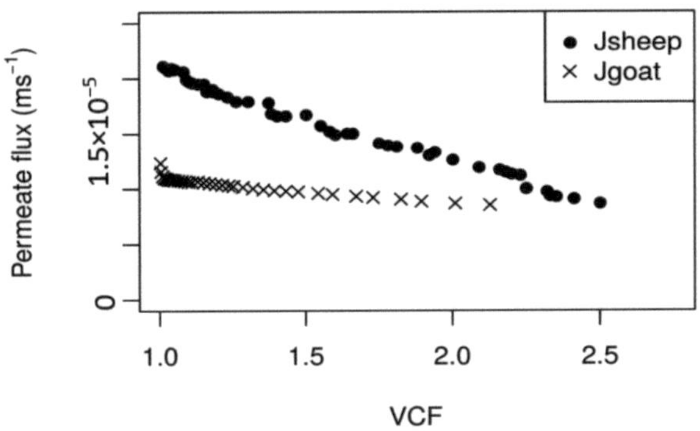

Figure 4.
Variation of average (three replicates) permeate fluxes with the volume concentration factor (VCF) for the concentration by nanofiltration of PUF-S ($\Delta P = 3.0 \times 10^6$ Pa; <v> = $1.42\ m\ s^{-1}$) and PUF-G ($\Delta P = 2.0 \times 10^6$ Pa; <v> = $0.94\ m\ s^{-1}$), at T = 25°C [20].

of lactose near the membrane surface, causing a higher increase in the osmotic pressure and concentration polarization phenomena. On the other hand, since the pH was 6.06 and the initial concentrations of calcium and phosphate were also higher than those of PUF-G, most probably, mineral fouling occurred due to the formation of insoluble calcium phosphates. In the case of PUF-G, the lower pH (5.43) and calcium and phosphate concentrations, due to the effect of dilution by diafiltration, were less prone to mineral fouling, leading to a more stable permeate flux. In spite of that, the permeate fluxes were lower during all the run, likely because of the highest concentration of chloride ions in goat cheese whey, which caused a greater initial osmotic pressure and therefore a lower effective transmembrane pressure. Beyond this, it is likely that also protein fouling contributed to this behavior since the pH of PUF-G was closest to the isoelectric point of β-lactoglobulin, the most abundant whey protein.

These results suggest that, in order to reach a better NF performance for recovering lactose, the following procedures should be applied: (1) precipitate calcium or use ionic exchange resins with the objective to reduce calcium concentration in the permeates, avoiding the decline of permeate fluxes during NF, due to the formation of calcium phosphates and (2) optimize NF/DF process to improve the performance of recovery process of lactose.

Membrane processes, for example, nanofiltration, also play a role in the recovery of mother liquor (or delactosed permeate) resulting from the crystallization process. This co-product was investigated after fractionation by membrane processes (NF and reverse osmosis) for salt substitute in soup formulations [26, 27]. By NF, the residual lactose was recovered and recycled to the crystallization tank, enhancing the yield of this process. On the other hand, the permeate was subjected to reverse osmosis producing a retentate enriched in salt, which will be used in the food industry. A detailed review about the possible valorization of the mother liquor is described by Oliveira and co-workers [28].

Several processes for lactose production involving advanced technologies are commercially available. Most of them include membrane processes, namely nanofiltration and reverse osmosis for the production of edible lactose, crystalline lactose, and lactose syrup, which can be used for the production of galactooligosaccharides.

The integration of membrane processes for recovering bioactive compounds from cheese whey, in small and medium dairies, in spite of the initial cost of the equipment, must be investigated in each case. The economic viability of these plants will depend on the valuation to be given to the different separated fractions. Cheese producer's associations may play a decisive role in the concentration of all the cheese whey released in a given region, in a single plant for processing/recovery of value-added compounds.

4. Conclusions

The recovery of lactose from cheese whey allows not only the valorization of this co-product in the cheese industry, but also to mitigate the environmental damage caused by it. This work is focused on the use of membrane processes for lactose recovery. The selection of the most suitable process depends on several factors such as composition of the initial cheese whey (quite varied, especially in the case of those resulting from artisanal cheese production), volume produced, and final intended application for lactose. Progress in these processes will lead to an overall improvement in the process of recovering lactose from cheese whey. In the case of membrane separation, its implementation at the industrial level is increasing. Hitherto, its use in small and medium scales is conditioned by the initial economic investment, depending rather on the synergy of the various producers, which in turn should be driven by their associations.

Acknowledgements

The author Antónia Macedo acknowledges to the project LACTIES (Innovation and eco-efficiency in the dairy sector), PDR2020-1.0.1.-FEADER-030773, and the research group LEAF (Linking Landscape, Environment, Agriculture and Food) for funding this work.

Conflict of interest

The authors declare no conflict of interest.

Author details

Antónia Teresa Zorro Nobre Macedo[1,2*], Joana Filipa Oliveira Monteiro[3],
David José Chaveiro da Silva Azedo[1],
Elizabeth da Costa Neves Fernandes de Almeida Duarte[2,3]
and Carlos Dias Pereira[4]

1 Higher Agriculture School, Polytechnic Institute of Beja, Beja, Portugal

2 Linking Landscape, Environment, Agriculture and Food (LEAF), Higher Institute of Agronomy, University of Lisbon, Lisbon, Portugal

3 Higher Institute of Agronomy, University of Lisbon, Lisbon, Portugal

4 Higher Agriculture School, Polytechnic Institute of Coimbra, Coimbra, Portugal

*Address all correspondence to: atmacedo@ipbeja.pt

References

[1] Chandrasekaran M. Valorisation of Food Processing By-products. Boca Raton: CRC Press; 2013. p. 771. ISBN: 978-1-4398-4887-6

[2] Walstra P, Wouters JTM, Geurts TJ. Dairy Science and Technology. 2nd ed. Boca Raton: CRC Taylor & Francis Group, LLC; 2006. p. 763. ISBN 0-8247-2763-0

[3] Daufin G, René F, Aimar P. Les separations par membrane dans les procédés de l'industrie alimentaire. Paris: Collection Sciences et Techniques Agroalimentaires; 1998. p. 588. ISBN: 978-2743002282

[4] Cheryan M. Ultrafiltration and Microfiltration Handbook. 2nd ed. Boca Raton: CRC Press LLC; 1998. p. 527. ISBN: 1-56676-598-6

[5] FAOSTAT. FAO Statistical Databases Cheese Whey [Internet]. 2017. Available from: http://www.fao.org/faostat/en/#data/BL [Accessed: 27 December 2017]

[6] European Whey Processes Association (EWPA). Economic Report 2016/17. 2017

[7] Mulder M. Basic Principles of Membrane Technology. 2nd ed. Dordrecht: Kluwer Academic Publishers; 2003. p. 564. ISBN: 0-7923-4248-8

[8] Lipnizki F. Cross-flow membrane applications in the food industry. In: Peinemann KV, Nunes SP, Giorno L, editors. Membrane Technology, Membranes for Food Applications. Vol. 3. Weinheim: Wiley-Vch Verlag GmbH& Co, kGaA; 2010. pp. 1-24. DOI: 10.1002/9783527631384.ch1

[9] Afonso CAM, Crespo JG. In: Afonso CAM, Crespo JG, editors. Green Separation Processes: Fundamentals and Applications. Michigan: Wiley; 2005. p. 363. ISBN: 3527309853, 9783527309856

[10] Vandezande P, Gevers LEM, Paul JS, Vankelecom IFJ, Jacobs PA. High throughput screening for rapid development of membranes and membrane processes. Journal of Membrane Science. 2005;**250**:305-310

[11] Mollea C, Marmo L, Bosco F. Valorisation of cheese whey, a by-product from the dairy industry. In: Muzzalupo I, editor. Food Industry. 2013:549-588. DOI: 10.5772/48737

[12] Duscher S. Ceramic membranes for the filtration of liquids: An actual overview. Highlights 2013. 2014;**14**: 13-21

[13] Macedo A, Duarte E, Fragoso R. Assessment of the performance of three ultrafiltration membranes for fractionation of ovine second cheese whey. International Dairy Journal. 2014; **2014**(48):31-37

[14] Macedo A, Duarte E, Pinho M. The role of concentration polarization in ultrafiltration of ovine cheese whey. Journal of Membrane Science. 2011;**381**: 34-40

[15] Vankelecom IFJ, Gevers LEM. Membrane processes. In: Afonso CAM, Crespo JG, editors. Green Separation Processes. Weinheim: Wiley-VCH Verlag GmbH &Co. KGaA; 2013. pp. 251-270. DOI: 10.1002/9783527631384.ch1

[16] Luo J, Ding L, Qi B, Jaffrin MY, Wan Y. A two-stage ultrafiltration and nanofiltration process for recycling dairy wastewater. Bioresource Technology. 2011;**102**:7437-7442

[17] Salehi F. Current and future applications for nanofiltration technology in the food processing. Food

and Bioproducts Processing. 2014;**92**: 161-177

[18] Suárez E, Lobo A, Alvarez S, Riera FA, Alvarez R. Partial demineralization of whey and milk ultrafiltration permeate by nanofiltration at pilot plant scale. Desalination. 2006;**198**:274-281

[19] Macedo A, Ochando Pulido JM, Fragoso R, Duarte E. The use and performance of nanofiltration membranes for agro-industrial effluents purification. In: Farrukh MA, editor. Nanofiltration. London: Intechopen Limited; 2018. pp. 65-84. DOI: 10.5772/intechopen.70909.ch4

[20] Macedo A, Monteiro J, Duarte E. A contribution for the valorisation of sheep and goat cheese whey through nanofiltration. Membranes. 2018;**8**:114. DOI: 10.3390/membranes8040114

[21] Guu YK, Zall RR. Nanofiltration concentration effect on the efficacy of lactose crystallization. Journal of Food Science. 2006;**57**(3):735-739

[22] Rice GS, Kentish SE, O'Connor AJ, Barber AR, Pihljamaki A, Nystrom M, et al. Analysis of separation and fouling behaviour during nanofiltration of dairy ultrafiltration permeates. Desalination. 2009;**236**:23-29

[23] Cuartas-Uribe B, Alcaina-Miranda SI, Soriano-Costa E, Mendoza-Roca JA, Iborra-Clar MI, Lora-García J. A study of the separation of lactose from whey ultrafiltration permeate using nanofiltration. Desalination. 2009;**241**: 244-255

[24] Feng Y, Chang X, Ma R. Separation of galactooligosaccharides mixture by nanofiltration. Journal of the Taiwan Institute of Chemical Engineers. 2009; **40**:326-332

[25] Bortoluzzi A, Faitão JA, Di Lucio M, Dallago RM, Steffens J, Zabot GL, et al. Dairy wastewater treatment using integrated membrane systems. Journal of Environmental Chemical Engineering. 2017;**5**:4819-4827

[26] Alexander LD, Stout MA, Drake M, Beckman SL, Metzger L. 0518 membrane fractionation of delactosed permeate to enhance salty taste. Journal of Animal Science. 2016;**94**(5):248-256

[27] Magueijo V, Minhalma M, Queiroz D, Geraldes V, Macedo A, Pinho MN. Reduction of wastewaters and valorisation of by-products from "Serpa" cheese manufacture using nanofiltration. Water Science & Technology. 2005;**52**(10–11):393-399

[28] Oliveira D, Puri R, Fenelon MA, O'Mahony JA. Delactosed permeate as a dairy processing co-product with major potential value. International Journal of Food Science and Technology. 2018:10. DOI: 10.1111/ijfs.14064

Chapter 4

Hundred Years of Lactitol: From Hydrogenation to Food Ingredient

Sergio I. Martinez-Monteagudo, Kaavya Rathnakumar, Maryam Enteshari, Collette Nyuydze, Juan C. Osorio-Arias and Hiran Ranaweera

Abstract

The first report on the synthesis of lactitol dates back to the early 1920s. Nearly 100 years have passed since then, and the applications of lactitol have exceeded its original purpose. Currently, lactitol is used in bakery, confectionery, chocolate, desserts, chewing gum, cryoprotectant, delivery agent, and stabilizer in biosensors. Lactitol is the main reaction product derived from the hydrogenation of lactose. This chapter is aimed at providing a succinct overview of the historical development of lactitol, a summary of its synthesis, and an overview of its properties and applications.

Keywords: lactitol, catalytic hydrogenation, sugar alcohols, low-calorie sweeteners, lactose utilization

1. Introduction

Lactitol, a sugar alcohol, is not found in nature, and its synthesis requires lactose in solution, hydrogen gas, and solid catalyst. The first attempts of lactitol synthesis were made about 100 years ago. Since then, the synthesis of lactitol has evolved into a highly efficient process with a projected production of 1.9 million metric tons by 2022 [1]. In a nutshell, the synthesis consists in the incorporation of a hydrogen ion into the carbonyl group of lactose. Such incorporation involves a set of multiple elementary reactions known as Langmuir-Hinshelwood-Hougen-Watson (LHHW) kinetics. The hydrogenation has consensually thought to occur by adsorption, reaction surface, and desorption of the reactants. A number of kinetics models suggest that the surface reaction is the predominant step [2]. Within the surface reaction, the reaction between two adsorbed species is catalyzed by a transition metal supported in an inert material. Over the years, several catalytic systems (metal and support) have been investigated in terms of their physical and chemical properties. An important feature of the catalytic hydrogenation is the multiphase nature of the reaction, where liquid, solid, and gas are brought into contact for a given time. Upon completion of the reaction, lactitol is separated from the slurry by centrifugation and crystallization. In crystalline form, lactitol can exist in four crystal forms, depending on the crystallization protocol [3]. Each type of crystal is characterized by its melting point and solubility. The most common structure of lactitol is the monohydrate form, and therefore it is the most studied one.

IntechOpen

Lactitol is best known as a nutritive sweetener, whose relative sweetness is between 30 and 40% comparable with that of sucrose [4]. More importantly, regulatory agencies such as European labeling and FDA consider a caloric value of lactitol as 2.4 and 2.0 kcal g^{-1}, respectively, which correspond to a reduction of 48–40% with respect to sucrose [5]. The molecular structure of lactitol offers stability over a wide range of pH and temperature, making it a suitable candidate for the synthesis of biopolymers, hydrogels, and surfactants. Over the last past decades, lactitol has emerged into a multipurpose ingredient from low-caloric sweetness to coating material in chewing gums.

This chapter summarizes relevant advancements over the 100 years of lactitol history. Section 2 provides a historical overview of lactitol, highlighting some of the most significant milestones. Section 3 discusses an overview of the catalysts used for the hydrogenation of lactose. Section 4 addresses some chemical and physical properties of lactitol. Finally, a summary of current and potential applications of lactitol is discussed in Section 5.

2. Historical timeline

Figure 1 illustrates selected milestones of lactitol over the past 100 years. A comprehensive review of the technological advancements of lactitol can be found elsewhere [1]. Chemical catalysis was perhaps the first great contributor to the advancement of lactitol. In 1920, Senderens [6] hydrogenated lactose over activated nickel. Senderens' catalysis was very unstable, making unrealistic any kind of large-scale production. The stability of nickel-based catalysts became a reality with the invention of the sponge nickel by Raney in 1925 and 1926 [7]. Raney's invention consisted of crystalline particles of active nickel embedded within an inactive metal. In subsequent years, the reaction kinetics of hydrogenation was elucidated, which allowed the production of lactitol at high yields and selectively.

Early production of lactitol was aimed at research facilities, where potential applications were investigated throughout 1930–1970. In 1938, the crystalline structure of lactitol was elucidated by purification and crystallization of the hydrogenated slurry [8]. A second anhydrous crystalline form of lactitol, dihydrate, was discovered by 1952 [9]. In subsequent years, lactitol entered the fields of nutrition, material science, and biotechnology. Fortification of infant food, synthesis of lactitol-based polyethers, sweetening agent, and animal feed are examples of applications of lactitol.

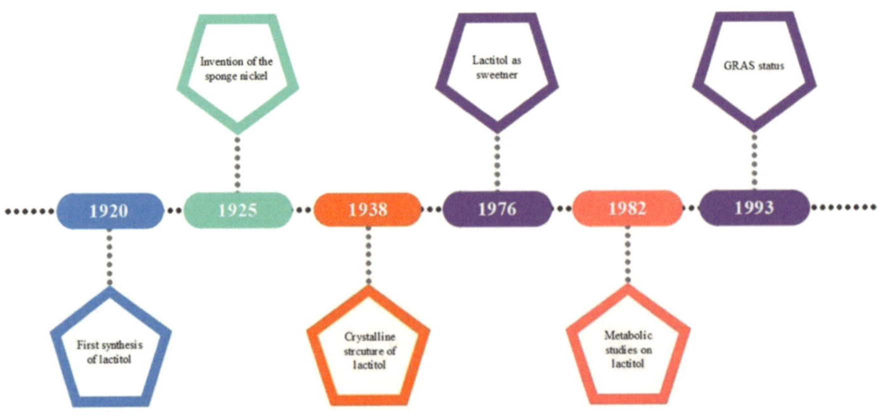

Figure 1.
Selected scientific and commercial milestones of lactitol over the past 100 years. Adapted from [1].

In 1977, the sweetness intensity of lactitol was established by the development of the sweetness scale using sucrose as a reference [10]. Soon after, lactitol was incorporated in confectionary formulations and chewing gum. In the 1980s, lactitol found applications in the field of hygiene and medicine, where it was used to formulate toothpaste, mouthwashes, and aseptic products. Metabolic concerns related to the consumption of lactitol were studied in 1981 [11]. In years thereafter, lactitol was used for the treatment of liver disease [12]. In 1993, the Food and Drug Administration (FDA) granted the status of Generally Recognized as Safe (GRAS) [13]. The current literature on applications of lactitol reveals about 3000 patents, ranging from a low-calorie sweetener to a surfactant and stabilizer agent. Nowadays, lactitol and other sugar alcohols represent a significant global market with various applications, and its production is projected to reach 1.9 million metric tons by 2022.

3. Production of lactitol

3.1 Catalytic hydrogenation

Lactitol is not found in nature, and it can only be produced through catalytic hydrogenation of lactose. Thus, the transition state theory of catalytic surface reactions is the foundation of lactitol synthesis. The actual synthesis consists of a sequence of elementary reactions, namely adsorption, surface reaction, and desorption [14]. Collectively, all these reactions are known as the Langmuir-Hinshelwood-Hougen-Watson (LHHW) kinetics [2]. **Figure 2** illustrates the LHHW kinetics that is formulated from a presumed elementary step. Then, the rate is derived through the different elementary steps with the assumption of one of them is the rate-determining step, while the others are achieved the equilibrium. The overall reaction rate is strongly affected by temperature and pressure since these variables determine the equilibrium of the elementary reactions.

3.1.1 Adsorption

Lactose and hydrogen are adsorbed through chemisorption, where the exchange of electrons with surface sites leads to the formation of a chemical bond [15]. Lactose is adsorbed from the bulk solution, a process that overcame the interaction forces of the solvent. A molecular mechanism is responsible for adsorbing the

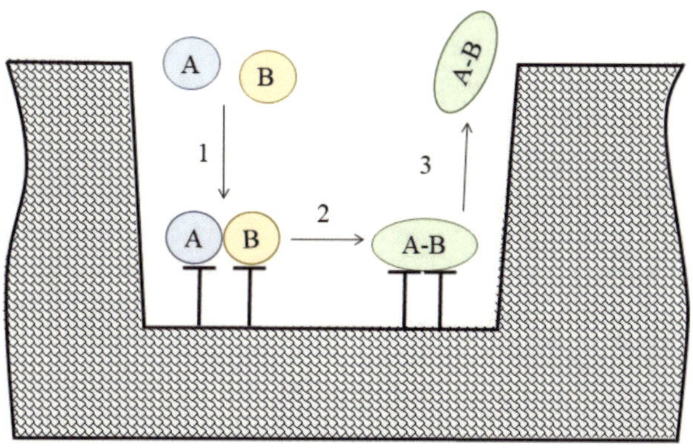

Figure 2.
Illustration of Langmuir-Hinshelwood-Hougen-Watson kinetic. Adapted from [2]. Numbers 1, 2, and 3 represent adsorption, surface reaction, and desorption, respectively.

lactose and hydrogen is followed a dissociative mechanism ($H_2 \leftrightarrow 2H^*$) due to the action of transition metals. Dissociative adsorption requires an adjacent vacant site, and the rate of attachment is proportional to the square of the vacant concentration [16]. The adsorption of reagents occurs within a very short timeframe. Once the adsorption is completed, the adsorbed molecules are in equilibrium with those molecules in the bulk phase.

3.1.2 Surface reaction

Examples of reaction mechanisms occurring at the surface include duel-site, single-site, two adsorbed species, and unabsorbed species [17]. Such mechanisms have been used for hydrogenation of a number of carbohydrates including, glucose, fructose, xylose, and lactose [18].

3.1.3 Desorption

The products of the surface reaction are subsequently desorbed into the reaction medium. Theoretically, the rate of desorption is exactly the opposite in sign to the rate of adsorption [19]. However, the desorption of reaction products is regarded as rapid and therefore neglected within the rate equation.

3.2 Catalysts

The design and selection of catalyst systems have been a major research topic in organic synthesis and chemical engineering. Several factors should be considered for the adequate selection of a catalyst system including, the transition metal, supporting material, preparation methods, and solvent. For lactose hydrogenation, metals such as nickel (Ni), ruthenium (Ru), and palladium (Pd) within a range of 1–10% are commonly used due to their relatively high reactivity and selectivity toward aldehyde groups. The concentration of the metal is linearly related to its activity within a limited range of 1 to 10%. Outside the concentration range, the metal is not available for reaction. A number of metal-based catalysts have been developed for lactitol production, including nickel-based, ruthenium-based, and other metal-based catalysts.

3.2.1 Nickel-based

Raney in 1920s patented a protocol where active metal (Ni) was embedded within an inactive metal (Al) frame [7]. The activity of the Raney's catalyst results from the random distribution of nickel crystals within the inactive crystal lattices [20]. A number of metals have been added into the Raney-nickel catalysts to further increase the reactivity [21]. Chromium, molybdenum, and tungsten are examples of metals added. The use of metal promoters (Cr-, Mo-, and Fe-Ni) showed a five-fold rate enhancement over non-promoted Raney nickel in the hydrogenation of glucose [21]. Although nickel-based catalysts are an effective catalyst for lactose hydrogenation, it suffers from the deactivation problem due to the nickel leaching and catalyst sintering. This results in a loss of catalyst activity and high nickel content in the lactitol product solution.

3.2.2 Ruthenium-based

Ruthenium is another metal used as a catalyst supported in different materials, such as carbon, alumina, silica, and synthetic gel. The ruthenium catalyst was effective for the hydrogenation of monosaccharides and disaccharides [22]. Ruthenium-based catalysts are supported in alumina (Ru/Al2O3), silica gel (Ru/gel), titanium dioxide (Ru/TiO2), crosslinked polystyrene (Ru/CP), and activated carbon (Ru/C). Ruthenium-based catalyst is more active than a nickel-based catalyst, and this leads to the higher catalyst selectivity [23]. Moreover, ruthenium-based catalyst was more stable than Ni based catalyst in the hydrogenation process; this leads to the extended lifetime of catalysts [23].

3.2.3 Other metal catalysts

Metals such as copper (Cu) and Pd have also been studied for the hydrogenation of lactose. For instance, Cu/SiO_2 was effective for the catalytic transformation of lactose to a high yield mixture (75–86%) of sorbitol and galactitol [24]. The boron nitride supported palladium (Pd/h-BN) was applied for the hydrogenation of lactose. The results indicated that the high lactose conversion ratio (up to 50%) was obtained with this catalyst.

3.3 Reaction pathways

Lactitol is the main product formed during the hydrogenation of lactose, followed by a considerable formation of lactulose, lactulitol, lactobionic acid, sorbitol, and galactitol [23]. All these compounds are formed through a combination of hydrogenation, isomerization, hydrolysis, and oxidation. **Figure 3** illustrates the reaction pathways occurring during the hydrogenation of lactose. A number of factors influence the occurrence and extent of a given reaction. Temperature, pressure, pH, agitation, type of catalysts, concentration, and catalyst load are examples of such factors [2].

3.3.1 Hydrogenation

Lactose is readily reduced to its corresponding alcohol, where the carbonyl group reacts with the hydrogen ion. This reaction is represented by scheme (1), and it is the main reaction occurring during the hydrogenation. Concomitantly, other reducing sugars (lactulose, galactose, and glucose) are also hydrogenated to form their respective alcohol (lactulitol, galactitol, and sorbitol). These reactions are exemplified in scheme (4), (10), and (9), respectively. Lactulose is formed through isomerization, while galactose and glucose are formed via hydrolysis of lactose.

3.3.2 Isomerization

The temperature used during hydrogenation may trigger the isomerization of lactose via enolization of the glucose molecule, scheme (4). Hypothetically, galactose, and glucose may undergo isomerization to form D-Tagatose and fructose, respectively. The yield corresponding to derives from isomerization is rather low. This is because the isomeric form of a reducing carbohydrate is prone to hydrogenation. Scheme (9) and (10) illustrate the hydrogenation of glucose and galactose, respectively.

3.3.3 Hydrolysis

Lactose may hydrolyze to some extent, leading to the formation of galactose and glucose, scheme (8). Lactulose and lactulitol may also hydrolyze, and their respective product can be hydrogenated. These sets of reactions are illustrated in

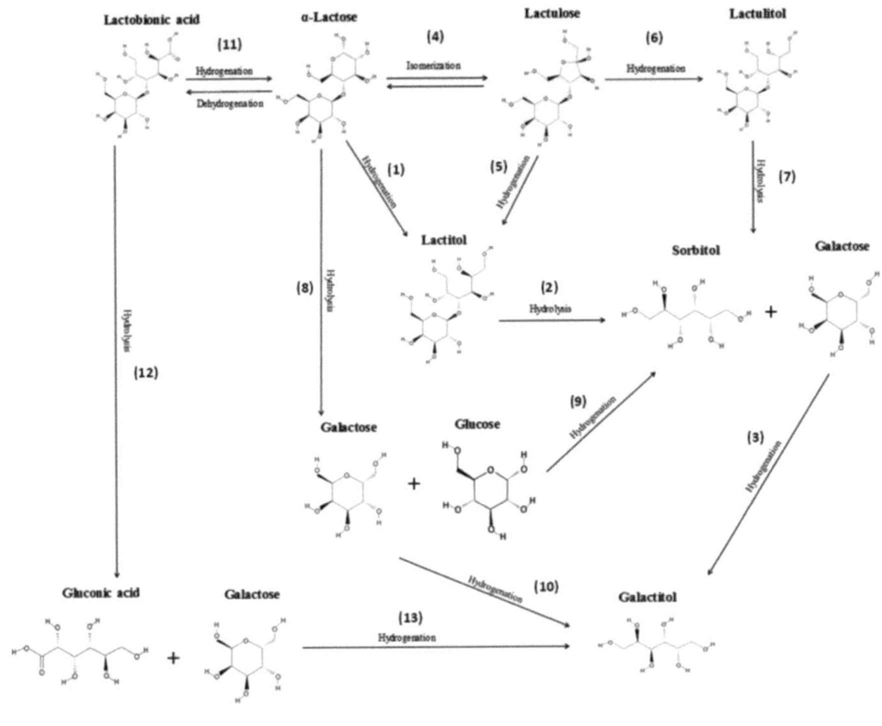

Figure 3.
Reaction scheme during catalytic hydrogenation of lactose.

scheme (5). Prolong reaction time and high temperature can induce the hydrolysis of lactitol, which leads to the formation of sorbitol and galactose, scheme (2).

3.3.4 Oxidation

Lactose can undergo dehydrogenation forming lactobionic acid, scheme (11). This situation would occur under a limited concentration of hydrogen. Subsequently, the lactobionic acid may undergo hydrolysis (scheme (13)) to form gluconic acid and galactose.

3.4 Production of lactitol

The industrial hydrogenation of lactose is commonly done in a batch mode using sponge nickel as a catalyst. In the hydrogenation of lactose, the reaction temperature ranged from 130 to 180°C, while the pressure of hydrogen gas varied between 50 and 170 bars. **Figure 4** exemplifies a hydrogenation batch reactor. The batch reactor is charged at the top of the tank. This type of reactor is based upon the movement of hydrogen from the gas phase to a liquid phase and across a liquid-solid interface to the surface of the supported catalyst, where the hydrogen gas is adsorbed. During the reaction, hydrogen gas is consumed by the catalytic reaction creating concentration gradients across the reactor. Such gradients control the net movement of hydrogen gas to the catalyst and, therefore, the speed of the reaction. Temperature, pressure, and agitation are the main variables controlling the reaction rate and final yield. Batch reactors offer the advantage of not having large temperature gradients, and the development of velocity profiles is negligible, which simplifies the operation. The performance of a batch reactor for catalytic hydrogenation depends on the hydrogen movement across the reactor. This feature is the principal disadvantage of the batch reactor since they are designed to control the mass transfer only through agitation.

Alternatively, catalytic hydrogenation can be performed by a continuous flow of reactants. Conceptually, the continuous operation has been exemplified in a trickle-bed reactor using structured catalysis [19]. A simplified diagram of the continuous hydrogenation of lactose is presented in **Figure 5**. The process mainly consists of feed streams, heat exchangers, reactor units, and separator. Kasehagen [25] exemplified the production of lactitol under continuous mode using a lactose solution (50% wt/wt) in water with sponge nickel (1.8%) at 160°C and 130 bar. Under such conditions, about 98% of lactose was converted into lactitol.

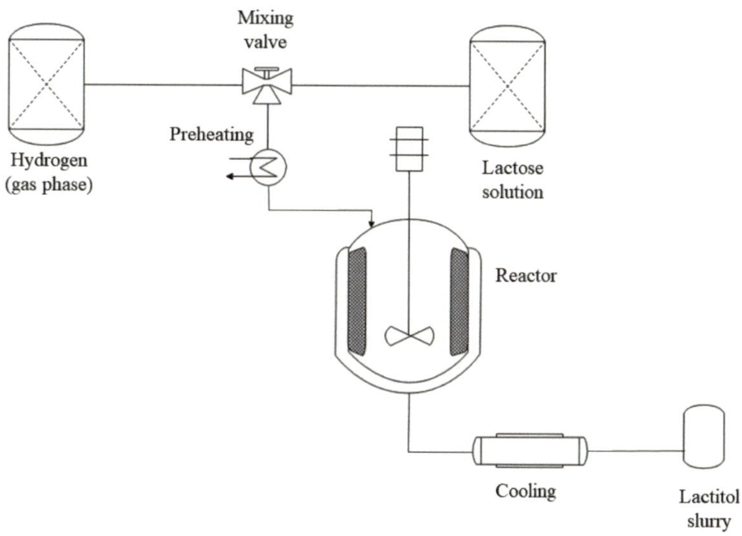

Figure 4.
Continuous-stirred tank reactor for batch hydrogenation of lactose.

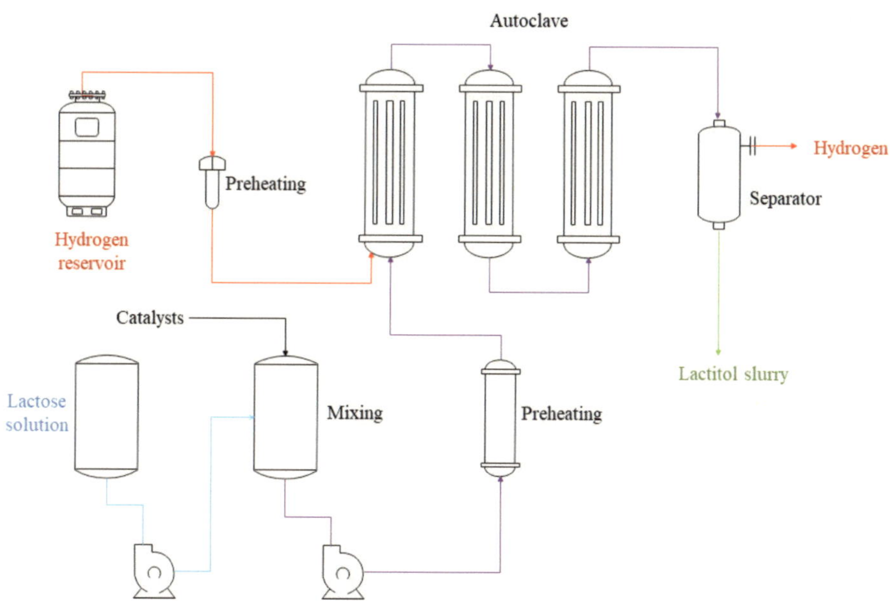

Figure 5.
Schematic representation of continuous hydrogenation of lactitol.

4. Properties of lactitol

4.1 Chemical and crystalline forms

In solid-state, lactitol can exist in different crystalline forms, having different melting points. Early observations showed the existence of two forms of anhydrous lactitol having different melting points [8, 9]. XRD and IR-spectra revealed three hydrate forms (mono-, di-, and tri-hydrate), two anhydrate (A and B), and one amorphous form [3, 26, 27]. The most common form of lactitol is monohydrate, which is obtained through slow crystallization of the lactitol slurry. Lactitol is a monoclinic polyol with one intra- and eight inter-molecular hydrogen bonds in its chemical structure [27]. All hydroxyl H-atoms form hydrogen bonds, which give rise to an eight-membered ring, chair conformation of the galactopyranosyl ring. The crystalline form of lactitol dihydrate is tetragonal with 3 intra- and 12 inter-molecular hydrogen bonds in its chemical structure [28]. Similar to lactitol monohydrate, all hydroxyl H-atoms form hydrogen bonds resulting in a chair configuration of the galactopyranosyl ring.

4.2 Solubility

Lactitol is found commercially as a crystalline powder. Interestingly, lactitol properties and therefore its potential application depend on the given crystalline form. Lactitol is recovered after hydrogenation, where the spent catalyst is removed via ion-exchange. Then, the slurry is evaporated under vacuum, and subsequently crystallized under prescribed protocol. Once the crystals are formed, the lactitol slurry is centrifugated and dried. Crystallization is the key step during the formation of a given crystal form. The crystallization of carbohydrates can be used as a general guideline of the crystallization of lactitol. Nurmi and Kaira [29] provided the most accurate guides in the literature for the crystallization of lactitol. Solubility curves of the lactitol crystals are illustrated in **Figure 6**. For simplicity, the solubility curve was divided into four regions.

The region IV illustrates the required conditions to yield lactitol in anhydrous form. This has been illustrated by Nurmi and Kaira [29], who crystallized a 91%

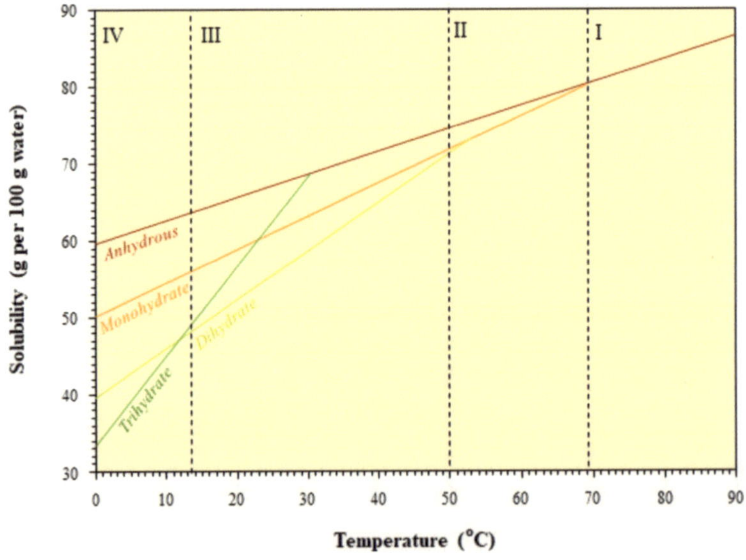

Figure 6.
Solubility curves of lactitol anhydrous, monohydrate, dihydrate, and trihydrate.

lactitol solution to obtain lactitol anhydrous. The solution was cooled from 95 to 75°C within 10 h, inducing the crystallization from solution. Similarly, Heikkila et al. [30] obtained lactitol anhydrous by cooling from 90 to 80°C a 90% lactitol solution. The working conditions for yielding lactitol monohydrate are exemplified in region II. Heikkilä et al. [30] obtained lactitol monohydrate using a four-step crystallization. Heikkilä's protocol involved the cooling of a 82% seeded lactitol solution from 70 to 40°C in 16 h. Wijnman et al. [31] obtained lactitol in the form of monohydrate by seeding a 80% solution of lactitol, and cooled it from 75 to 50°C in 18 h. The remaining mother liquid from this protocol was seeded and cooled down to 18–15°C to obtain lactitol dihydrate, indicated in region II. Wijnman et al. [31] followed a similar protocol to obtain 60% yield of lactitol dihyrate. Lactitol trihydrate, which is illustrated in region I, is obtained by further crystallization of the mother liquid at temperatures lower than 10°C [29].

4.3 Caloric value

Evidence of the reduced-calorie value of lactitol dates back to the 1930s, where the enzymatic hydrolysis of lactitol was found to be significantly slower than that of lactose [32]. This observation pointed out the possibility of a reduced calorie effect of lactitol. Indeed, Hayashibara and Sugimoto [33] measured the concentration of lactitol in the intestines of rabbits injected with a 20% solution of lactitol. After hours of injection, the lactitol concentration did not, while the concentration of glucose was reduced by 85%. Van Es et al. [34] analyzed the metabolized energy derived from lactitol and sucrose. They found that the energy contribution to the body was 60% less than for sucrose. European labeling considers a blanket caloric value of lactitol as 2.4 kcal g^{-1} [5]. At the same time, the Food and Drug Administration (FDA) establishes a general value of 2.0 kcal g^{-1}, a reduction of 48–40% with respect to sucrose.

4.4 Sweetness

Lactitol is known for its mild and clean sweet taste [4]. Relative sweetness is measured in relation to a reference value of 1, which corresponds to the sucrose sweetness at a given concentration [5]. Lactitol possess a relative sweetness from 0.3 to 0.42. Generally, lactitol sweetness is considered to be 30–35% of the sucrose sweetness. Thus, simply replacing sucrose with lactitol requires substantial amount of lactitol. Alternatively, lactitol is combined with other sweeteners to synergistically reduce the sucrose concentration.

4.5 Health claims

Lactitol is not considered as essential nutrient, but its consumption has been clinically linked to a number of health benefits. Health benefits and claims associated with the consumption of sugar alcohols have been reviewed elsewhere [35, 36]. Overall, sugar alcohols are a limited source of energy for oral bacteria that results in less production of acid. van der Hoeven [37] studied the cariogenicity of lactitol in fed rats, and observed that lactitol significantly reduced caries development when compared with sucrose. This observation was in agreement with the rate of fermentation by oral bacteria. Acid production from lactitol occurred at much lower rate than the acid production of sucrose. Clinical evidence demonstrated a reduction in the incidence of caries by the substitution of sucrose with sugar alcohols in chewing gum and candies. van Loveren [38] postulated that the preventive effects against caries in gums and candies formulated with sugar alcohols are due to a stimulation of the salivary flow, providing a buffer capacity that washes away soluble carbohydrates. However, there is no consensus regarding the minimal dose required to reduce caries. Nevertheless, van Loveren [38] suggested that chewing of sugar-free chewing gum at least 3 times per day may reduce caries incidence.

Lactitol is frequently prescribed as a laxative agent for the treatment of chronic constipation [39]. As a laxative agent, lactitol is minimally absorbed in the small intestine, and when it reaches the large intestine, it creates an osmotic gradient that increases the water retention in the stool, enhancing its passage. Miller et al. [40] performed a meta-analysis on the efficacy and tolerance of lactitol for adult constipation. It was found that lactitol supplementation was not only well tolerated but also significantly improved symptoms of constipation.

5. Applications of lactitol

5.1 Cryoprotectant and dryoprotectant

Lactitol is a polyol having the ability of preventing physical and chemical degradation of protein preparations during frozen and drying. The effectiveness of lactitol as a cryoprotectant agent was demonstrated in fish muscle (rainbow trout), where lactitol preserved the structure of myofibrillar proteins [41]. Interestingly, lactitol influenced the kinetic of formation of hydrophobic residues in the surface of proteins. Similarly, Nopianti et al. [42] added lactitol to prevent protein denaturation of threadfin bream surimi during 6 months of frozen storage. A formulation made of 6% of lactitol resulted in protective effect comparable with that obtained for polydextrose and sorbitol. Ramadhan et al. [43] cryoprotected duck surimi by the addition of lactitol. More importantly, the studied by Ramadhan et al. [43] showed a protective effect after five cycles of freeze-thaw during 4-month of frozen storage.

Lactitol can form glassy matrices within the protein structure that immobilizes the system and preventing unfolding. Moreover, lactitol may form hydrogen bonds with the surrounding protein, helping the preservation the enzymes. Such mechanisms have been validated during the drying of protein preparations. Kadoya et al. [44] freeze-dried a solution of L-lactic deydrogenase and bovine serum albumin using lactitol monohydrate as a cryoprotective agent. Microscopic observation indicated the formation of hydrogen bonds that substitute water molecules, and maintaining the activity of L-lactic dehydrogenase. This is an important observation showing the protective effect of lactitol in pharmaceutical applications that helps to minimize product immunogenicity.

The preservation of archeological artifacts has benefited from the protective effect of lactitol. The stability of archeological wood was performed by the impregnation of lactitol prior to freeze-drying. It was showed that the impregnation of lactitol resulted in higher hygroscopicity compared with polyethylene glycol impregnation [45]. Babiński [46] treated waterlogged archeological oak with lactitol, and evaluated changes in dimensions and moisture content. Lactitol reduced the wood shrinkage after freeze-dried by replacing water molecules and fill the cell walls.

5.2 Surfactant and hydrogel

The structure of lactitol confers higher chemically stability than lactose and sucrose. Lactitol stability is due to the absence of the carbonyl group, resulting stability over a broad range of pH (3–9). Moreover, lactitol is not a reducing sugar (absence of carbonyl group) which does not participate in the Maillard reactions. Such properties of lactitol offer potential for non-conventional applications, such as surfactants, emulsifiers, and hydrogels. Indeed, Van Velthuijsen [47] produced a non-ionic emulsifier made of lactitol via esterification of palmitic acid under alkaline conditions. Lactitol esters displayed relevant detergent activity by removing soil and stains from towels. Dupuy et al. [48] determined the micellization of lactitol-based surfactants in water. It was found that lactitol surfactants were barely dispersed at low concentrations, and the formation of micelles was due to their stearic hindrance. Drummond and Wells [49] produced mono-esters of lactitol with chain lengths from C8 to C16—octyl, dodecyl, and hexadecyl. The interfacial tension of such surfactants was determined by putting them in contact with hexadecane and triolein. The chain of the surfactant minimally reduced the interfacial tension than their shorter chain counterpart. Surfactants made of lactitol displayed the tendency to foaming over 30 min. This is an important observation indicating the great potential of lactitol based surfactants to be used as emulsifiers. It is worth to mention that surfactants made of lactitol have not been produced commercially.

Disaccharides from renewable sources can be used as building blocks for the synthesis of polymers and hydrogels. Wilson et al. [50] produced polyether polyols via lactitol propoxylation at alkaline environment. Lactitol polyether polyols showed similar viscosity and hygroscopicity than their counterpart sucrose-based polyols of the same hydroxyl number. Moreover, the decomposition of lactitol polyols was negligible. Wilson et al. [50] prepared rigid polyurethane foams from lactitol polyether polyols. Lactitol based foams showed physical properties comparable to that of the commercial foams. Hu et al. [51] hydrogenated sweet whey permeates and synthesized polyurethane foams by propoxylation of lactitol slurry. The lactitol foams were showed low-density, strong mechanical properties, and thermal stability. Lin et al. [52] controlled the propoxylation of lactitol to produce polyether polyols with nine polypropylene oxide branches. Such lactitol polyether polyols were used to prepare hydrogel via acylated polyethylene glycol bis carboxymethyl ether. Lactitol hydrogels absorbed water up to 1000% of their dry weight. Remarkably, these hydrogels expelled free water at a temperature above 30°C.

Lactitol can be seeing as building block compound to design delivery systems for bioactive compounds. Already, Han et al. [53] prepared poly(ether polyol) hydrogel from lactitol, and it was showed ability of delivering acetylsalicylic acid over a pH range of 4–9. More importantly, the release was controlled by the amount crosslinking of the hydrogel. Han et al. [53] used lactitol cross-linked hydrogel to incorporate protein for controlled release of the protein into the surrounding fluid. It was found that the release of β-lactoglobulin, bovine serum albumin, and γ-globulin was constant over 2 h in a temperature range of 37–45°C. Constant release at such temperature range approaches the human body temperature, suggesting the use of lactitol based delivery system for clinical applications. Chacon et al. [54] prepared hydrogels of lactitol having swelling capacity up to 81-fold. The length of polypropylene oxide branches and the extent of crosslinking controlled the swelling capacity of the hydrogels. Chacon et al. [54] added a lipase within the lactitol hydrogel for temperature-controlled release. About 90% of the enzyme was released into the medium within the first 60 min at temperatures between 25 to 40°C. The development of drug delivery systems used lactitol as a target group [55], where the carrier is incorporated in liposomes for treatment of liver disease.

5.3 Bakery

Sugar reduction and replacing in bakery formulations has not been a trivial task in the past. This is because sugar not only provides a pleasant taste but also plays a critical role in the development of the quality characteristics of the batter or dough. Psimouli and Oreopoulou [56] replaced sugar with lactitol in equal amount for cake formulations. The resulting batter was comparable in terms of flow index and the temperature of starch gelatinization. Sensory analysis indicated no significant difference between the batter formulated with lactitol and the one formulated with sugar. Frye and Setser [57] employed lactitol as a sweetener to optimize cake formulations having a reduction of 45% in the caloric content. Such formulations showed comparable attributes with a standard layer cake. Similarly, Zoulias et al. [58] evaluated the role of lactitol and other polyols as a sucrose replacement on the texture profile of cookie dough. The lactitol formulated dough resulted in medium values of hardness and consistency.

5.4 Chocolate and confectionary

The formulation of sugar-free chocolate represents a significant challenge because the entire sugar needs to be replaced, which in turns, affects the melting properties of the chocolate [59]. Mentink and Serpelloni [60] formulated a low-calorie chocolate having an equimolar blend of maltitol, lactitol, and isomaltulose. The formulation showed technical and organoleptic properties comparable to those of traditional formulation with sucrose. Synergistic effects have been reported when sugar alcohols are combined with other sweeteners. de Melo et al. [61] developed a sugar-free chocolate having acceptable sensory scores by the combination of high-intensity sweeteners and blends of sugar alcohols.

Sugar alcohols have also been used in the manufacture of hard-boiled sweets. Blends of lactitol, sorbitol, and mannitol provided sticky texture due to their hygroscopic nature. Such challenge is the principal limitation in the formulation of hard-boiled candies with sugar alcohols. Serpelloni and Ribadeau-Dumas [62] enhanced the process of hard coating by using a syrup of sugar alcohols. Another investigation on the role of replacing sugar in syrups demonstrated that about 40% of the total sugar can be replaced with lactitol without changes in the moisture content and density [63]. Lactitol addition produced a two-fold increase in the viscosity of the syrup. Blankers et al. [64] formulated a syrup sweetening suitable for soft confectionery applications. The syrup is made of lactitol and polydextrose, and it is combined with the lactitol slurry derived from lactose hydrogenation.

5.5 Chewing gum

Lactitol in combination with other sugar alcohols is used to formulate sugar-free chewing gum. The hygroscopicity of lactitol is relatively low, which facilities its incorporation into the gum. Huzinec et al. [65] incorporated lactitol within the microcrystalline cellulose carrier. With such blend, the release of flavor was extended in chewing gums. McGrew et al. [66] used active compounds in combination with mannitol, xylitol, maltitol, lactitol, and hydrogenated starch hydrolysates to control release of such active agent that are embedded in the gum base. Yatka et al. [67] formulated a generic gum base containing oligofructose and sorbitol, maltitol, xylitol, lactitol, and mannitol. Such a generic formulation was blended with glycerol. Subsequently evaporated to produce a low-moisture and sugar-free chewing gum. The combination of oligofructose and sugar alcohols improved quality properties, including texture, moisture adsorption. Reed et al. [68] formulated hard-coated chewing gum coated with a layer of lactitol, maltitol, and sorbitol.

5.6 Biosensor development

Lactitol can be used as an additive for biosensors because of the stabilizing effect on enzymes. Karamitros and Labrou [69] used lactitol to immobilize isoenzyme glutathione transferase. About 5% of lactitol resulted in a prolonged stability of the enzymes. Gibson and Woodward [70] combined diethylaminoethyl-dextran hydrochloride (DEAE-Dextran) and lactitol for the stabilization of enzymes in a dry state. Such combination of DEAE-Dextran (10%) and lactitol (5%) preserved up to 95% of the activity after 16 d. Zhybak et al. [71] immobilized creatinine deaminase and urease in the presence of lactitol, and reported improvement in the stability of the biosensor. Remarkably, biosensor selectivity was not impacted by the addition of lactitol.

6. Conclusions

Over the past 100 years, lactitol has been evolving successfully finding new applications while its original purpose has expanded. Today, lactitol is added into a number of food formulations, such as bakery, confectionery, chocolate, desserts, chewing gum, and cryoprotectant. Research strategies for expanding the applicability of lactitol are needed including, solubility at different conditions, rheological behavior, heat stability, thermogravimetric analysis, stability toward heat and pH, particle size, bulk, and particle density, and crystallization kinetics.

Acknowledgements

This work has been made possible through the financial support of USDA National Institute for Food and Agriculture (HATCH projectSD00H607-16).

Conflict of interest

The authors declare no conflict of interest.

Author details

Sergio I. Martinez-Monteagudo[1,2,3*], Kaavya Rathnakumar[1], Maryam Enteshari[1], Collette Nyuydze[1], Juan C. Osorio-Arias[4] and Hiran Ranaweera[1]

1 Dairy and Food Science Department, South Dakota State University, Brookings, SD, United States

2 Family and Consumer Sciences, New Mexico State University, Las Cruces, NM, United States

3 Chemical and Materials Engineering, New Mexico State University, Las Cruces, NM, United States

4 Aoxlab Research Group, Development and Innovation Department, Aoxlab S.A.S., Medellin, Colombia

*Address all correspondence to: sergiomm@nmsu.edu

References

[1] Martínez-Monteagudo SI, Enteshari M, Metzger L. Lactitol: Production, properties, and applications. Trends in Food Science & Technology. 2019;**83**:181-191

[2] Cheng S, Martínez-Monteagudo SI. Hydrogenation of lactose for the production of lactitol. Asia-Pacific Journal of Chemical Engineering. 2019;**14**(1):e2275

[3] Yajima K, Okahira A, Hoshino M. Transformation of lactitol crystals and dehydration with grinding. Chemical & Pharmaceutical Bulletin. 1997;**45**(10):1677-1682

[4] Zacharis C, Stowell J. Lactitol. In: O'Brien-Nabors L, editor. Alternative Sweeteners. Boca Raton, FL: CRC Press; 2011. pp. 316-326

[5] Radeloff MA, Beck RHF. Polyols – More than Sweeteners. Zuckerindustrie. Sugar Industry. Vol. 1382013. pp. 226-234

[6] Senderens JB. Catalytic hydrogenation of lactose. Comptes Rendus. 1920;**170**:47-50

[7] Raney M. Method of Preparing Catalytic Material. United States: Patent and Trademark Office; 1925. US1563587A

[8] Wolfrom ML et al. Crystalline lactositol. Journal of the American Chemical Society. 1938;**60**:571-573

[9] Wolfrom ML, Hann RM, Hudson CS. Lactitol dihydrate. Journal of the American Chemical Society. 1952;**74**:1105

[10] Lee C-K. Structural functions of taste in the sugar series: Taste properties of sugar alcohols and related compounds. Food Chemistry. 1977;**2**(2):95-105

[11] Wiel-Wetzels WAM. Metabolic consequences of the use of polyalcohols and fructose—A literature review, part 1: Xylitol. Voeding. 1981;**42**(3):78-81

[12] Booij CJ. Lactitol for the Treatment of Liver Disease. U.K.: Intellectual Peroperty Office; 1983

[13] Food and Drug Administration, PURAC Biochem b.v. Filing of Petition for Affirmation of GRAS Status (Lactitol) Federal Register. 1993. Retrieved from: https://www.gpo.gov/fdsys/granule/FR-1994-08-05/94-19098

[14] Froment GF, Bischoff KB, De Wilde JD. Chemical Reactor Analysis and Design. New Jersey, United States: John Wiley & Sons, Inc.; 2011

[15] Brahme PH, Doraiswamy LK. Modelling of a slurry reaction. Hydrogenation of glucose on Raney nickel. Industrial and Engineering Chemistry Process Design and Development. 1976;**15**(1):130-137

[16] Crezee E et al. Three-phase hydrogenation of d-glucose over a carbon supported ruthenium catalyst—Mass transfer and kinetics. Applied Catalysis A: General. 2003;**251**(1):1-17

[17] Wisnlak J, Simon R. Hydrogenation of glucose, fructose, and their mixtures. Industrial & Engineering Chemistry Product Research and Development. 1979;**18**(1):50-57

[18] Mikkola J-P, Salmi T, Sjöholm R. Modelling of kinetics and mass transfer in the hydrogenation of xylose over Raney nickel catalyst. Journal of Chemical Technology & Biotechnology. 1999;**74**(7):655-662

[19] Déchamp N et al. Kinetics of glucose hydrogenation in a trickle-bed reactor. Catalysis Today. 1995;**24**(1):29-34

[20] Raney M. Method of Producing Finaly-Divided Nickel. United States; 1927. US5665406A

[21] Gallezot P et al. Glucose hydrogenation on promoted Raney-nickel catalysts. Journal of Catalysis. 1994;**146**(1):93-102

[22] Gilman Boyer G. Hydrogenation of Mono-and Disaccharides to Polyols. United States; 1994. US2868847A

[23] Doluda VY et al. Kinetics of lactose hydrogenation over ruthenium nanoparticles in hypercrosslinked polystyrene. Industrial & Engineering Chemistry Research. 2013;**52**(39):14066-14080

[24] Zaccheria F et al. Catalytic upgrading of lactose: A rest raw material from the dairy industry. Green Chemistry. 2017;**19**(8):1904-1910

[25] Kasehagen L. Continuous Hydrogenation of Sugars. U.S.: Patent and Trademark Office; 1953

[26] Kanters JA, Schouten A, van Bommel M. Structure of lactitol (4-O-[beta]-d-galactopyranosyl-d-glucitol) monohydrate: An artificial sweetener. Acta Crystallographica Section C. 1990;**46**(12):2408-2411

[27] Kivikoski J et al. Crystal structure of lactitol (4-O-β-d-galactopyranosyl-d-glucitol). Carbohydrate Research. 1992;**223**:45-51

[28] Kivikoski J, Valkonen J, Nurmi J. Crystal structure of lactitol (4-O-β-D-galactopyranosyl-D-glucitol) dihydrate. Carbohydrate Research. 1992;**223**:53-59

[29] Nurmi J, Kaira M. Process for the Crystallization of Lactitol. U.S.: Patent and Trademark Office; 2002

[30] Heikkilä H et al. Crystallization of Lactitol, Crystalline Lactitol Product and Use Thereof. U.S.: Patent and Trademark Office; 2002

[31] Wijnman CF, Van Velthuijsen JA, Van Den Berg H. Patent and Trademark Office. United States; 1998. US5726303A

[32] Karrer P, Buchi J. Reduction products of disaccharides: Maltitol, lactitol and cellobitol. Helvetica Chimica Acta. 1937;**20**:86-90

[33] Hayashibara K, Sugimoto K. Containing Lactitol as a Sweetener. U.S.: Patent and Trademark Office; 1976

[34] Van Es AJH, De Groot L, Vogt JE. Energy balances of eight volunteers fed on diets supplemented with either lactito1 or saccharose. British Journal of Nutrition. 1986;**56**(3):545-554

[35] Livesey G. Health potential of polyols as sugar replacers, with emphasis on low glycaemic properties. Nutrition Research Reviews. 2003;**16**(2):163-191

[36] Nath A et al. Biochemical activities of lactose-derived prebiotics—A review. Acta Alimentaria. 2017;**46**(4):449-456

[37] van der Hoeven JS. Cariogenicity of lactitol in program-fed rats (short communication). Caries Research. 1986;**20**(5):441-443

[38] van Loveren C. Sugar alcohols: What is the evidence for caries-preventive and caries-therapeutic effects? Caries Research. 2004;**38**(3):286-293

[39] Prasad VGM, Abraham P. Management of chronic constipation in patients with diabetes mellitus. Indian Journal of Gastroenterology. 2017;**36**(1):11-22

[40] Miller LE, Tennilä J, Ouwehand AC. Efficacy and tolerance of lactitol supplementation for adult constipation: A systematic review and meta-analysis. Clinical and Experimental Gastroenterology. 2014;7:241-248

[41] Herrera JR, Mackie IM. Cryoprotection of frozen-stored

actomyosin of farmed rainbow trout (*Oncorhynchus mykiss*) by some sugars and polyols. Food Chemistry. 2004;**84**(1):91-97

[42] Nopianti R et al. Cryoprotective effect of low-sweetness additives on protein denaturation of threadfin bream surimi (*Nemipterus* spp.) during frozen storage. CyTA Journal of Food. 2012;**10**(3):243-250

[43] Ramadhan K, Huda N, Ahmad R. Freeze-thaw stability of duck surimi-like materials with different cryoprotectants added. Poultry Science. 2012;**91**(7):1703-1708

[44] Kadoya S et al. Freeze-drying of proteins with glass-forming oligosaccharide-derived sugar alcohols. International Journal of Pharmaceutics. 2010;**389**(1):107-113

[45] Majka J, Babiński L, Olek W. Sorption isotherms of waterlogged subfossil scots pine wood impregnated with a lactitol and trehalose mixture. Holzforschung. 2017:813

[46] Babiński L. Dimensional changes of waterlogged archaeological hardwoods pre-treated with aqueous mixtures of lactitol/trehalose and mannitol/trehalose before freeze-drying. Journal of Cultural Heritage. 2015;**16**(6):876-882

[47] Van Velthuijsen JA. Food additives derived from lactose: Lactitol and lactitol palmitate. Journal of Agricultural and Food Chemistry. 1979;**27**(4):680-686

[48] Dupuy C et al. Influence of structure of polar head on the micellization of lactose-based surfactants. Small-angle X-ray and neutron scattering study. Langmuir. 1998;**14**(1):91-98

[49] Drummond CJ, Wells D. Nonionic lactose and lactitol based surfactants: Comparison of some physico-chemical properties. Colloids and Surfaces a-Physicochemical and Engineering Aspects. 1998;**141**(1):131-142

[50] Wilson M et al. Preparation and characterization of lactitol-based poly(ether polyol)s for rigid polyurethane foam. Journal of Applied Polymer Science. 1996;**59**(11):1759-1768

[51] Hu M et al. Polyurethane rigid foam derived from reduced sweet whey permeate. Journal of Agricultural and Food Chemistry. 1997;**45**(10):4156-4161

[52] Lin W et al. Thermosensitive lactitol-based polyether polyol (LPEP) hydrogels. Journal of Polymer Science Part A: Polymer Chemistry. 1998;**36**(6):979-984

[53] Han JH et al. Lactitol-based poly(ether polyol) hydrogels for controlled release chemical and drug delivery systems. Journal of Agricultural and Food Chemistry. 2000;**48**(11):5278-5282

[54] Chacon D et al. Swelling and protein absorption/desorption of thermo-sensitive lactitol-based polyether polyol hydrogels. Polymer. 2000;**41**(23):8257-8262

[55] Luo L-H et al. Pharmacokinetics and tissue distribution of docetaxel liposome mediated by a novel galactosylated cholesterol derivatives synthesized by lipase-catalyzed esterification in non-aqueous phase. Drug Delivery. 2016;**23**(4):1282-1290

[56] Psimouli V, Oreopoulou V. The effect of alternative sweeteners on batter rheology and cake properties. Journal of the Science of Food and Agriculture. 2012;**92**(1):99-105

[57] Frye AM, Setser CS. Optimizing texture of reduced-calorie yellow layer cakes. Cereal Chemistry. 1992;**69**(3):338-343

[58] Zoulias EI, Oreopoulou V, Kounalaki E. Effect of fat and sugar replacement on cookie properties. Journal of the Science of Food and Agriculture. 2002;**82**(14):1637-1644

[59] Aido RP, Depypere F, Afoakwa EO, Dewettinck K. Industrial manufacture of sugar-free chocolates—Applicability of alternative sweeteners and carbohydrate polymers as raw materials in product development. Trends in Food Science & Technology. 2013;**32**(2):84-96

[60] Mentink L, Serpelloni M. U.S.: Patent and Trademark Office; 1994

[61] de Melo L, Bolini HMA, Efraim P. Sensory profile, acceptability, and their relationship for diabetic/reduced calorie chocolates. Food Quality and Preference. 2009;**20**(2):138-143

[62] Serpelloni M, Ribadeau-Dumas G. United States: Patent and Trademark Office; 1995. US08531646

[63] Onwulat CI, Konstance RP, Holsinger VH. Properties of reduced-fat composites of sugar alcohols, whey isolates and pectin. Journal of Food Lipids. 2000;**7**(1):39-50

[64] Blankers I et al. Patent and Trademark Office. United States; 2002. US6444250B1

[65] Huzinec RJ, Kearns TR, Schindeldecker TL. Comestible Products Having Extended Release of Additives and Method of Making. United States; 1999. US5912030A

[66] McGrew GN et al. Nutraceuticals or Nutritional Supplements and Method of Making. U.S: Patent and Trademark Office; 2005

[67] Yatka RJ, Richey LC, Meyers MA. Chewing Gum Products Using Oligofructose. Patent and Trademark Office; 1995. WO1993012666A1

[68] Reed MA et al. Hard Coated Chewing Gum with Improved Shelf Life, with Xylitol and Polyol Coatings. United States; 1994. US5376389A

[69] Karamitros CS, Labrou NE. Preserving enzymatic activity and enhancing biochemical stability of glutathione transferase by soluble additives under free and tethered conditions. Biotechnology and Applied Biochemistry. 2017;**64**(5):754-764

[70] Gibson TD, Woodward JR. Enzyme Stabilization. United States; 1993. US5240843A

[71] Zhybak M et al. Creatinine and urea biosensors based on a novel ammonium ion-selective copper-polyaniline nano-composite. Biosensors and Bioelectronics. 2016;**77**:505-511

Chapter 5

Bioconversion of Lactose from Cheese Whey to Organic Acids

José Manuel Pais-Chanfrau, Jimmy Núñez-Pérez, Rosario del Carmen Espin-Valladares, Marcos Vinicio Lara-Fiallos and Luis Enrique Trujillo-Toledo

Abstract

Organic acids constitute a group of organic compounds that find multiple applications in the food, cosmetic, pharmaceutical, and chemical industries. For this reason, the market for these products is continuously growing. Traditionally, most organic acids have been produced by chemical synthesis from oil derivatives. However, the irreversible depletion of oil has led us to pay attention to other primary sources as possible raw materials to produce organic acids. The microbial production of organic acids from lactose could be a valid, economical, and sustainable alternative to guarantee the sustained demand for organic acids. Considering that lactose is a by-product of the dairy industry, this review describes different procedures to obtain organic acids from lactose by using microbial bioprocesses.

Keywords: lactose, cheese whey, organic acids, acetic acid, lactic acid, citric acid, L-ascorbic acid, succinic acid, propionic acid, butyric acid, hyaluronic acid

1. Introduction

Organic acids (OAs) are compounds with relatively weak acidity properties [1, 2]. Carboxylic acids with one or more carboxyl groups (–COOH) are the most common OAs, following the sulfonic acids (–SO_2OH). Under certain circumstances, alcohol (with a group –OH) can also act as acid. Other groups, like thiol (–SH), enol, and phenol, also can confer acidity character to solutions, but all of them are very weakly acidic. Nowadays, many industrially produced organic acids (OAs) are synthesized from nonrenewable sources like petroleum oil [3]. Still, as can be expected, these sources could be depleted shortly, and it would lead to finding new renewable sources to produce OAs [4, 5].

Among others, a promising raw material is agro-industrial wastes (AIWs) [6, 7]. By its nature, AIWs could classify as complex organic compounds, which include mono- and polysaccharides, fats, and proteins. These raw materials are biotransforming by microbes in nature, so it is also able to metabolize AIWs into several OAs. Some of AIWs are by their constitution liquids like cheese whey (CW), molasses; but others are solids like bagasse, and citrus, potato, and banana peels. For liquid AIWs, the submerged fermentation (SmF), anaerobic or aerobic, is a suitable alternative [8–10], while solids could use the solid-state fermentation (SSF) [8, 11–13].

Some revisions regarding the microbial production of OAs have been published [3, 14–16]. Also, some authors focused their attention on the use of AIWs in SSF to produce OAs [11–13, 17–19].

Volatile fatty acids (VFAs) are the smallest and simplest organic acids [20]. VFAs can be classified as short-chain fatty acids (SCFA, C_2-C_6 carboxylic acids), medium-chain fatty acids (MCFAs, C_7-C_{12}), long-chain fatty acids (LCFA, C_{13}-C_{21}), and very-long-chain fatty acids (C_{22} and higher) [21, 22]. SCFAs and MSFAs are commonly involved in the anabolic process and in the energy metabolism of mammalian cells. SCFAs are produced by colonic bacteria and are metabolized by the liver and enterocytes, whereas MCFAs are gotten from triglycerides that are found, for example, in milk or dairy products [23, 24]. OAs have been used since time immemorial by humankind in the seasoning of foods and sauces, such as vinegar, and more recently has been widely used as food additives, preservatives, descaling and cleaning agents [3, 25, 26]. They can also be used as precursors of other more complex organic compounds of broad utility in fine and pharmaceutical chemistry [27, 28].

OAs have certain relevant usefulness characteristics like its preservative, buffering and chelating capacity, in addition to its traditional use as an acidulant in food formulations, and most of them are GRAS classified [9, 28]. Among others, the foremost OAs are citric, acetic, lactic, tartaric, malic, gluconic, ascorbic, propionic, acrylic, and hyaluronic acids [28]. Nowadays, citric acid is the most widely produced OA in the world [29, 30].

The preferred carbon source to achieve their biosynthesis is glucose. Other sugars like fructose, galactose, maltose, and cellobiose can be metabolized for many bacteria and yeast. While cellulose, lignin, and more complex polysaccharides could be adequately transformed by using fungi [31], in this review, however, are mainly discussed the different reports showing that lactose also can be used to produce organic carboxylic acids with different uses.

2. The cheese whey and lactose

Lactose ($C_{12}H_{22}O_{11}$, MW 342.297 g mol^{-1}, IUPAC name: β-D-galacto-pyranosyl-(1 → 4)-D-glucose) is a disaccharide present naturally in milk and dairy products [32]. Today lactose is produced mainly as sweet whey from cheese-making industry as a by-product [33]. Lactose contents in whole milk are 4.9% for cows, and 4.8% for sheep and goats [34]. Water (94% wt.), lactose (4.5% wt.), protein (0.6% wt.), mineral salts (0.35% wt.), ash (0.5% wt.), and some traces of fat (500 ppm) and lactic acid (500 ppm) are the main components in sweet whey [35].

There are numerous technologies for the processing of the whey generated from the production of the various types of cheese [36–39]. Almost all start with pasteurization of cheese whey (CW) to decrease the microbial bioburden and to reduce the degradation of lactose and whey proteins. Subsequently, solid–liquid separation stages are usually used to remove the casein micro-lumps and the fat that may still contain the CW, using clarifying and disk centrifuges, for this purpose [40].

The defatted and pasteurized CW can then be subjected to microfiltration to retain the bacteria debris, before proceeding to the separation of the proteins, lactose, and mineral salts [41]. Membrane filtration has been used to isolate the whey proteins, mineral salts, and water present in CW [38, 42–44]. In this sense, ultrafiltration membranes can be suitable to isolate whey proteins, while nano-filters can separate the remaining lactose and mineral salts. Finally, the separated products are usually concentrated using evaporators, and dried, using technologies such as spray drying (SD) [45–48].

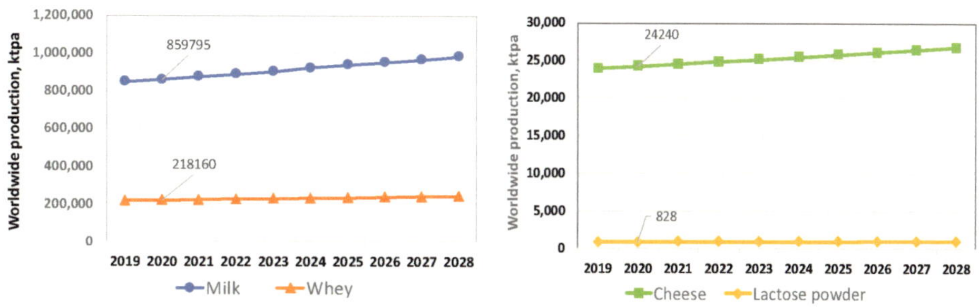

Figure 1.
Worldwide production projections (in metric kilo-tonnes per annum) of milk, cheese, whey, and lactose powder up to 2028 [51].

The most valuable components of whey are, in this order, whey proteins, lactose, and mineral salts. From a conventional process of obtaining lactose from sweet whey, whey powder (on March/2020, 880 EUR/ton), as well as deproteinized whey powder, lactose powder, mineral salts powder, and powder of whey proteins, can be obtained. From the latter, which is the product with the highest added value (on March/2020, 2030EUR/ton), different whey proteins presentations are usually obtained, like whey protein concentrate (WPC), whey protein hydrolysates (WPH), and whey protein isolate (WPI) [49].

As worldwide milk and cheese production has seen a constant increment in recent years, several millions of tons of whey are produced annually as a by-product [50] (**Figure 1**). A significant portion of whey has been used as an animal [51] and human feed supplementation due to its content of value proteins and minerals [52–56]. However, the enormous volumes of whey generated often overcome in many places the capacity of dairy-waste treatment plants [57]. For this reason, have focused the attention of numerous researchers' intent to valorize the whey and diminish the quantity of whey treated as waste [57–61].

Additionally, lactose is the component of whey that most contributes to the high biochemical oxygen demand (BOD) and chemical oxygen demand (COD) values in the dairy wastes [58, 62–64], bringing values around 30–50 and 60–80 kg m^{-3}, respectively [58, 64]. The great volumes of whey generated in the dairy industry could be the main obstacle to the further growth of cheese production in the next years [57]. One of the direct ways to reduce the adverse effects on the environment exerted by whey is using lactose containing the whey [65]. Lactose or "milk sugar" is a disaccharide formed by galactose and glucose, has a sweetening power, slightly lower than sucrose [32, 66]. It is usually used as a food additive [33, 67] or as a starting raw material for other products of agro-industrial interest [68, 69].

3. Organic acid market: overview and perspectives

The citric acid (2415 kilo-tonne per annum (ktpa)), L-ascorbic acid (132 ktpa), tartaric acid (30 ktpa), itaconic acid (43 ktpa), and bio-acetic acid (1830 ktpa) were produced by microbial fermentation, while gluconic acid (50 ktpa, with a 67:33 proportion between fermentative and chemical synthesis way), lactic acid (35 ktpa, 50:50), and malic acid (30 ktpa, 30:70) were produced by both fermentation and chemical synthesis, and, finally, some organic acids, like formic acid (1150 ktpa), butyric acid (80 ktpa), propionic acid (50 ktpa), and fumaric acid (20 ktpa) were chemically synthesized [12, 70–72]. This outlook and its proportions have not changed much today, and the global market of OAs shows a sustainable growth of 5.48% AAGR (average annual growth rate) in the last years and it is expected that it could increase globally up to US$9.29 billion by 2021 and US$11.39 billion by 2022 [3, 73–75].

Biosynthesis of an OA is obtained by the biochemical pathway of cellular metabolism, as the final end product or as an intermediate product of a path [26]. Bacteria and fungi are the most available and suitable living organisms for the industrial production of OAs. The microbial production of organic acids is usually an attractive route for industrial implementation compared to chemical synthesis because the conditions used in microbial bioprocesses tend to be less extreme (in terms of temperature, pressure, extreme pH) and more friendly to the environment [3, 76]. However, this may be effective only if the concentration of these acids in the fermentation broth are high enough (in the order of tens or hundreds of grams per liter), and these are obtained in reasonably short times [77]. Also, the microbial bioconversion of sugars into organic acids is frequently carried out by strict anaerobic microorganisms, with relatively long fermentation, reduced productivity, and low titers of organic acids in the fermentation broth [27]. Those facts conspire with its large-scale implementation, and to turn the biotechnology in an economically attractive choice to the production of organic acids (**Figure 2**) [3, 26, 78].

In this context, the processes of isolation and purification of organic acids become critical [78, 79]. Various alternatives for the isolation and purification of organic acids from fermentation broth or biomass have been used. Among the most used primary purification methods are precipitation with Ca-salt or hydroxide [77], ammonium salt, organic solvents [80], and ionic solutions [81]. Microbial fermentation can produce directly only a few organic acids [74], and even more scarce are the microorganisms that can use lactose to achieve this.

3.1 Acetic acid

Acetic acid ($C_2H_4O_2$, MW 60.052 g mol^{-1}, IUPAC name: Ethanoic acid) is a monocarboxylic acid commonly used as a chemical starting reagent in the production of important chemicals, like cellulose acetate, polyvinyl acetate, and synthetic fibers. Vinegar (near 4% vol. acetic acid) is produced by fermentation of different carbon sources by acetic acid bacteria [82] and is widely employed in food preparation and cooking since ancient times. Currently, three-quarters of the world production is obtained by carbonylation of methanol (by chemical synthesis), basically from nonrenewable sources, while 10% is still obtained from the microbial biotransformation of sugars [83]. By 2014, the global acetic acid market reached 12,100 ktpa, with an average price of US$ 550 per ton and average annual growth of 4–5% [14]. In 2018, world production reached 16,300 ktpa, near to US$ 12.48 billion, forecasting production of 20,300 ktpa by 2024. China with 54% and the US (18%) are the largest producers [84, 85].

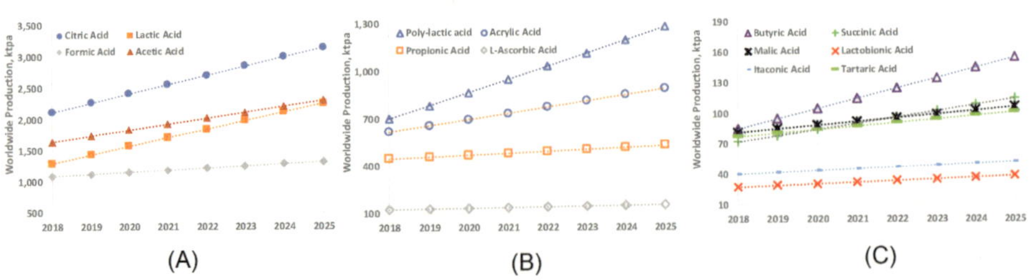

Figure 2.
Worldwide production of some organic acids between 2018 and 2025. (A) High-, (B) medium -, and (C) low-level of global production.

3.2 L-ascorbic acid

A case is the L-ascorbic acid (vitamin C, $C_6H_8O_6$, MW 176.124 g mol^{-1}, IUPAC Name: (5R)-[(1S)-1,2-Dihydroxyethyl]-3,4-dihydroxyfuran-2(5H)-one), one of the organic acids with the highest production and sales volumes today. Vitamin C can be obtained by microbial biosynthesis but from D-sorbitol [86].

Ascorbic acid, previously called hexuronic acid, is a soluble white solid and organic compound that presents itself as two enantiomers: L-ascorbic acid (vitamin C), and D-ascorbic acid, without any biological role found [87, 88]. Vitamin C is an essential nutrient for humans and many animals, and its deficiency can cause scurvy, in the past a common disease among sailors in long sea voyages [89]. It is used in as a food additive and a dietary supplement for its antioxidant properties [87, 88]. There is a report, however, that achieves the synthesis of vitamin C from the lactose present in the cheese whey, but through a defined group of chemical reactions [90]. In 2015 was produced 150.2 ktpa of ascorbic acid with a revenue of US$820.4 million. By 2017, China produced near to 95% of the world supply of vitamin C, having revenue of US$880 million [91].

3.3 Butyric acid

Butyric acid ($C_4H_8O_2$, MW 88.106 g mol^{-1}, IUPAC Name: Butanoic acid) is a mono-carboxylic acid, and it is an oily, colorless liquid that is soluble in water, ethanol, and ether. Salts of butyric acid are known as butyrates. Butyric acid is a chemical, commonly used as a precursor to produces other substances, like biofuel [92, 93], cellulose acetate [94, 95], and methyl butyrates [96], the two last coatings, and flavors compounds, respectively. Chemical synthesis is still the primary way of production of butyric acid due to the availability of raw material [92]. But some research explores the microbial biotransformation from renewable sources like agro-industrial wastes [72]. *Clostridium tyrobutyricum* can produce butyric acid from lactose, present in milk and cheese, along with H_2, CO_2, and acetic acid [97]. By 2016, the butyric acid worldwide market was around 80 ktpa, with a price of US$ 1800 per ton [98]. By 2020, global production of butyric acid is expected to reach 105 ktpa [99].

3.4 Citric acid

Citric acid ($C_6H_8O_7$, MW 192.123 g mol^{-1}, IUPAC Name: 2-Hydroxy-propane-1,2,3-tri-carboxylic acid) is a water-soluble tricarboxylic acid. Citric acid is widely used in the food and pharmaceutical industry due to its antimicrobial, antioxidant, and acidulant properties [100]. Citric acid can be produced from the citrus (like lemon, orange, lime, etc.), by chemical synthesis, or microbial fermentation [101]. Many microorganisms have been used to produce citric acid by microbial fermentation [102–104]. Among others, the fungus *Aspergillus niger* is the preferred choice to produce several useful enzymes and metabolites due to its ease of handling, and it being able to achieve high yields by using different cheaper agricultural by-products and wastes [101, 105]. By 2018, the worldwide citric acid production was more than 2000 ktpa, more than a half was produced in China. The global citric acid market is projected to reach a level of around 3000 ktpa by 2024, growing at a 4% CAGR during this period.

3.5 Propionic acid

Propionic acid ($C_3H_6O_2$, MW 74.079 g mol^{-1}, IUPAC Name: Propanoic acid) is an organic acid, colorless oily liquid with an unpleasant smell. Propionic acid

(PA) is a valuable mono-carboxylic acid used in chemical, pharmaceutical, and food industries, as a mold inhibitor, as a preservative of foods, as a significant element in the vitamin E production, and as a chemical intermediate in the chemical synthesis of cellulose fiber, perfumes, herbicides, etc. [16, 106, 107]. Today, propionate is mainly obtained for two processes. From ethylene, a nonrenewable source synthesized from oil, through the Reppe process [108], or from ethanol and carbon monoxide catalyzed by boron trifluoride (by the Larson process) [109].

Although chemical synthesis is the primary way of its production, the microbial production of PA is gaining attention and importance due to the depletion of petroleum sources and due to pieces of evidence of the more environmentally friendly microbial process [107, 110]. Propionibacterium is the most employed microorganism used for PA large-scale production [107, 111]. In 2020, the worldwide production of PA would reach 470 ktpa. The leading producers remain to be in Germany (BASF SE), USA (Dow Chemical Co. and Eastman Chemical), and Sweden (Perstorp). At the same time, the primary consumers are in the EU, USA, China, and India.

3.6 Lactic acid

Lactic acid ($C_3H_6O_3$, MW 90.078 g mol^{-1}, IUPAC Name: 2-Hydroxypropanoic acid) was the first organic acid commercially produced by microbial fermentation [112]. Bacterial fermentation of carbohydrates had been the main way for the industrial production of lactic acid (LA) with production level between 70 and 90% for 2009 [113]. The rest of production was achieved by chemical synthesis mainly from acetaldehyde coming from crude oil [114]. A racemic mixture of LA commonly is obtained by chemical synthesis, while L-lactic acid can be obtained by homofermentative anaerobic bacteria like *Lactobacillus casei* and *Lactococcus lactis*. Otherwise, heterofermentative bacteria produced carbon dioxide, ethanol, and/or acetic acid in addition to LA [115].

LA is currently used and has been approved as a food additive, preservative, decontaminant, and flavoring agent (with a code E270) [116, 117]. Also, it is used for chemical synthesis [118], mainly to produce poly-lactic acid (PLA), a thermal- and bioplastic polyester with widespread use in many applications [119, 120]. PLA is used, for example, in medical implants [121], as plastic fiber material in 3D-printing [122, 123], and as a decomposable packing material [124, 125].

In 2020, LA and PLA worldwide production will be around 1571 and 800 ktpa, respectively, with China, USA, EU, and Japan being the primary producers [126].

3.7 Succinic acid

Succinic acid ($C_4H_6O_4$, MW 118.088 g mol^{-1}, IUPAC Name: Butanedioic acid) has been widely used in many industries, as a food, detergent, and toner additive, for solders and fluxes, and as an intermediary commodity in the chemical and pharma industry [127]. After the increment of oil prices and diminishing availability of nonrenewable sources, researchers turned their attention over to the renewable feedstocks to produce succinic acid. SA as an intermediate in many biochemical pathways could be produced by many microorganisms and use many carbon sources [127]. For instance, the anaerobic-facultative bacteria *Actinobacillus succinogenes* can produce succinic acid from sugar cane molasses alone [128] or supplement with corn steep liquor powder [129].

Glucose as a carbon source has also been used to produce succinic acid by engineering strains of *Corynebacterium glutamicum* [130], *Escherichia coli* [131], and

Saccharomyces cerevisiae [132]. Succinic acid (SA) is a bulk OA commodity, and by 2010 the bioproduction was between 16 and 30 ktpa, and its expected annual growth was 10% [133], and by 2025, it is expected to exceed 115 ktpa [134].

3.8 Other acids

No reports of microbial obtention of tartaric ($C_4H_6O_6$, dicarboxylic acid), itaconic ($C_5H_6O_4$, dicarboxylic acid), and fumaric acid ($C_4H_4O_4$, dicarboxylic acid) from lactose have been found. Some of those, however, can be obtained indirectly, since there are published studies of the biosynthesis of itaconic acid [135–137], fumaric acid [138, 139] from glucose, and the latter can be obtained from the chemical or enzymatic hydrolysis of lactose.

4. Microbial bioprocesses for obtaining organic acids based on lactose

Like other renewable sources based on residual plant biomass from agricultural productions rich in complex polysaccharides, lactose has been used as a starting raw material to establish bioprocesses to produce different organic acids. Although there are microbial enzymes capable of breaking the bonds of polysaccharides, this would involve energy and time, which in the case of lactose would be less complicated and faster. In the case of lactose, this could become the starting material for

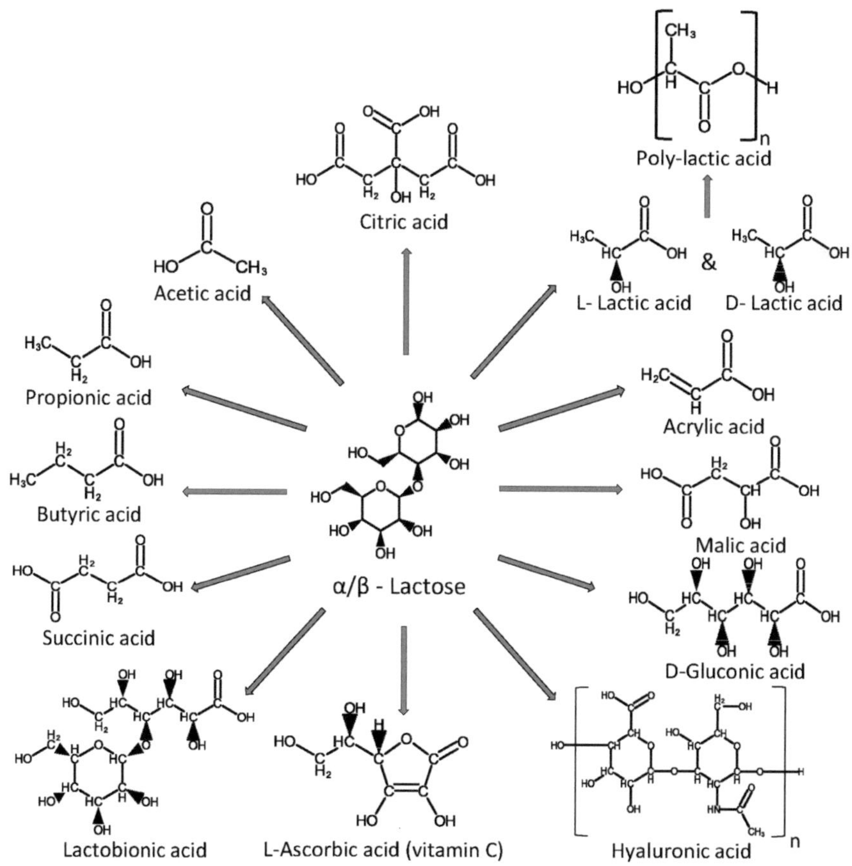

Figure 3.
Some of the organic acids that can be obtained microbially from lactose or whey.

Name	Source	Microorganism(s)	Culture conditions and production results*	Ref.
Acetic acid, $C_2H_4O_2$	WP	*Clostridium thermolacticum* and *Moorella thermoautotrophica*	Anaerobic, batch, 58°C, pH 7.2, 300 h, 0.81 g g^{-1}, 98 mM	[140, 141]
	WP	*Acetobacter aceti*	Aerobic, continuous membrane bioreactor, at 303 K, D = 0.141 h^{-1}, 96.9 g L^{-1}, 0.98 g g^{-1}, 4.82 g L^{-1} h^{-1}	[142–144]
	CW	*Propionibacterium acidipropionici*	Anaerobic, batch, 35°C, pH 6.5, 78 h, 0.11 g L^{-1} acetic acid + 0.33 g L^{-1} propionic acid	[145]
	CW	*Lactobacillus acidophilus*	Anaerobic, 37°C, 72 h, pH 6.5, 7 g L^{-1}	[146]
Acrylic acid, $C_3H_4O_2$	SCW	*Clostridium propionicum*	Anaerobic, +propanoic and acetic acids, 33°C, pH 7.1, 0.133 mmol g^{-1}	[147]
L-Ascorbic acid, $C_6H_8O_6$	CW	*Kluyveromyces lactis*	Aerobic, shake-flask, 48 h, 30°C, 30 mg L^{-1}	[148]
	Gal	*Saccharomyces cerevisiae* *Zygosaccharomyces bailii*	Aerobic, shake-flask,144 h, 30°C,0.40 g g^{-1}, 70 mg L^{-1}	[149]
Propionic acid, $C_3H_6O_2$	SWP	*Propionibacterium acidipropionici*	Anaerobic, fibrous bed bio-reactor (immobilized cells), 135 ± 6.5 g L^{-1}	[150]
	CW	*P. acidipropionici*	Anaerobic facultative, 6.1 g L^{-1}	[151]
		Propionibacterium freudenreichii	Anaerobic, 8.2 g L^{-1}	
	CW	*P. acidipropionici*	Anaerobic, 0.33 g L^{-1}	[145]
Lactic acid, $C_3H_6O_3$	SWP	*Lactobacillus casei*	Anaerobic, 36 h, pH 6.5, 37°C, 33.73 g L^{-1}	[152]
	SWP	*Lactobacillus rhamnosus*	Anaerobic, 37°C, pH 6.2, 200 rpm, 50 h, 143.7 g L^{-1}	[153]
	CW	*Lactobacillus acidophilus*	Anaerobic, 37°C, 72 h, pH 6.5, 42.62 g L^{-1}	[146]
	CW	Mixed culture of acetogenic and fermentative bacteria	Dark anaerobic, 35°C, HDT = 1 day, 10.6 g L^{-1} day^{-1}	[154]
Butyric acid, $C_4H_8O_2$	CW	*Clostridium beijerinckii*	Anaerobic, 37°C, pH 5.5, 0.08 g L^{-1} h^{-1}, 12 g L^{-1}	[155]
	CW	*Clostridium butyricum*	Anaerobic, + 5 g L^{-1}YE or + 50 μg L^{-1} biotin, 37°C, pH 6.5, 19 g L^{-1}	[156]
Succinic acid, $C_4H_6O_4$	CW	*Anaerobiospirillum succiniciproducens*	Anaerobic+CO_2, + Glu, pH 6.5, 39°C, 36 h, 16.5 g L^{-1}, 0.33 g L^{-1} h^{-1}	[157]
	CW	*Actinobacillus succinogenes*	Anaerobic+CO_2, 38°C, pH 6.8, 48 h,28 g L^{-1}, 0.44 g L^{-1} h^{-1}	[158]
	PWP	*Enterobacter* sp. LU1	Microaerobic, + Gly, 34°C, pH 7, 288 h, 69 g L^{-1}	[159]

Name	Source	Microorganism(s)	Culture conditions and production results*	Ref.
Malic acid, $C_4H_6O_5$	Milk	*Escherichia coli* K-12	Stationary culture for 72 h at 37°C, 168 mg g^{-1} DW	[160]
Gluconic acid, $C_6H_{12}O_7$	CW	*Pseudomonas taetrolens*	Aerobic, + Glu, 30°C, aeration: 1 L min^{-1}, 350–500 rpm, pH 6.5, 8.8 g L^{-1}	[161]
Citric acid, $C_6H_8O_7$	CW	*Aspergillus niger* ATCC9642	Aerobic, +15% sucrose, 30°C, 16 h, 106 g L^{-1}	[162]
Lactobionic acid, $C_{12}H_{22}O_{12}$	CW	*Pseudomonas taetrolens*	Aerobic, 30 °C, + Gly, aeration: 1 L min^{-1}, 350–500 rpm, pH 6.5, 78 g L^{-1}	[161]
	CW		Aerobic, + Lac, 30°C, aeration: 1 L min^{-1}, 350–500 rpm, pH 6.5, 100 g L^{-1}	
Hyaluronic acid, $(C_{14}H_{21}NO_{11})_n$	CCW, HCW	*Streptococcus zooepidemicus*	Aerobic (1 vvm), 37°C, pH 6.7 and 500 rpm	[163]
	Lac	*Lactococcus lactis*	Anaerobic, 1% Lac + 10 ng mL^{-1} nisin, 30°C, 24 h (12 h after induction), 0.6 g L^{-1}	[164]

WP: whey permeate; PWP: powder whey permeate; CW: cheese whey; SCW: sweet cheese whey; SWP: sweet whey powder; CCW: concentrate cheese whey; HCW: hydrolysate cheese whey; Gal: galactose; Gly: glycerol; Glu: glucose; Lac: lactose; YE: yeast extract; HDT: hydraulic detection time.
**In terms of concentration, yield, and/or productivity of the acid.*

Table 1.
Characteristics of some organic acids produced by bioconversion of lactose from commercial products or agro-industrial by-products.

the production by microbial bioprocesses, not only of the most demanded organic acids today but of other less-used ones that still not as highly in demand. However, subsequent studies must be carried out to make these technologies a viable and economically attractive alternative [3, 19].

Nowadays, however, some organic acids can be obtained by microbial bioprocesses directly from lactose (**Figure 3**), cheese whey, or both, using the different routes of their metabolisms (**Table 1**). The most demanded organic acids, like citric, acetic, and lactic acids, have been produced from whey (**Table 1**). Even more complex organic acids like poly-lactic and hyaluronic acids can also be produced from lactose. Another advantage of microbial production is related to the possibility of producing the racemic biological active acids exclusively. L-lactic acid is produced almost exclusively by lactic-acid bacterium *Lactobacillus casei* or L-ascorbic acid (vitamin C) by certain recombinant yeast strains of *Kluyveromyces lactis* or *Saccharomyces cerevisiae* [148, 149]. However, for some of the organic acids, the titers reached are still too low for these bioprocesses to be scaled to industrial production in an economically feasible way, and the chemical synthesis remains the most desired choice. At the industrial scale, to produce organic acids competitively, it would be necessary to have adequate sources of *raw materials* (cheap and renewable) and enhanced *microbial strains* (easy and safe to handle and able to work at high productivity). Also, it would be necessary to dispose of industrial facilities and technical expertise (*technical constrains*) to achieve it (**Figure 4**).

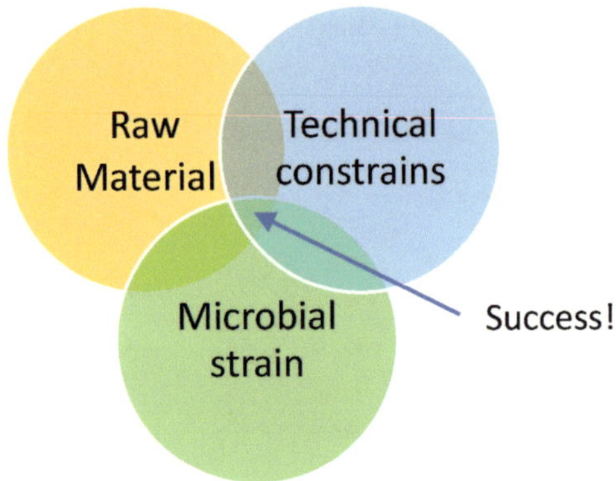

Figure 4.
Successful commercial production of organic acids by microbial biotransformations: keys to success.

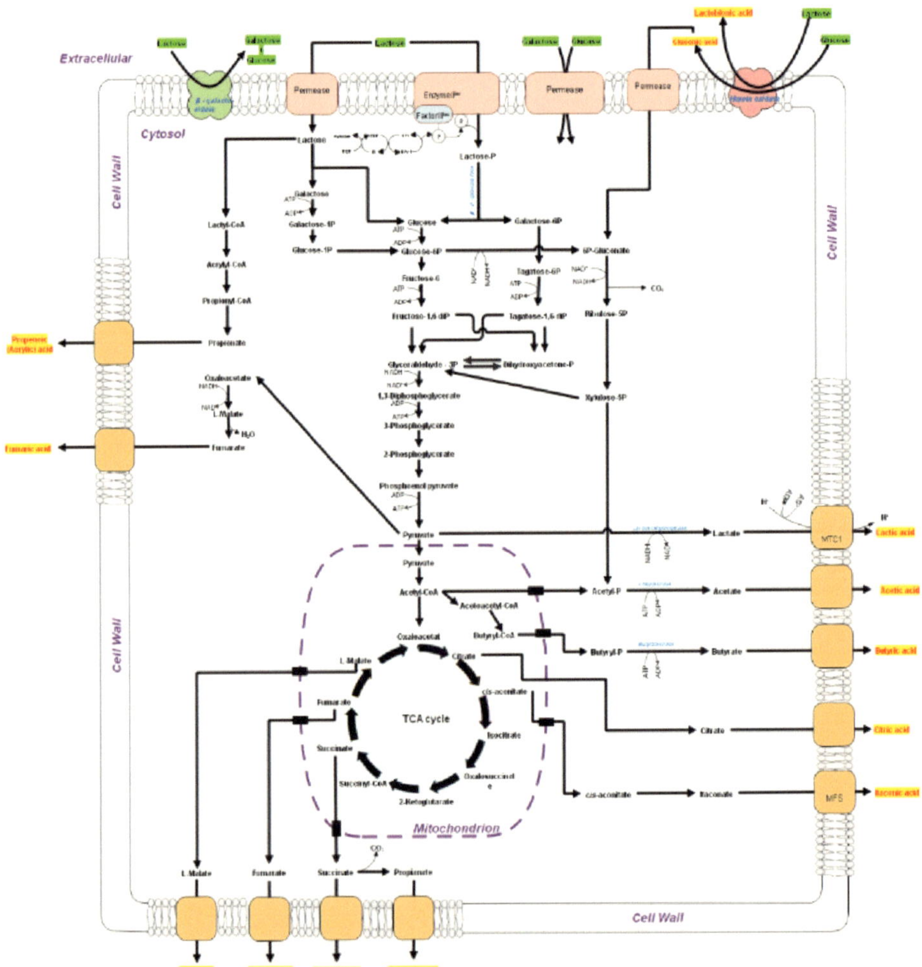

Figure 5.
Some of the microbial metabolic pathways for the synthesis of organic acids.

The microbial bioprocesses could be enhanced through optimization of up- and downstream processes that must be combined with metabolic engineering to increase productivity. Also, genetic engineering techniques could be used to obtain robust industrial strains that raise the expression levels of the genes involved in the metabolic pathways of synthesis of organic acids or repress others that deviate to produce unwanted by-products [164, 165].

Some of the identified metabolic pathways are associated with the tricarboxylic acid (TCA) cycle and demonstrate that most organic acids represent metabolites associated or partially associated with growth (**Figure 5**). A detailed study of these pathways can address the overexpression of some genes or repression of others using genetic engineering techniques.

5. Conclusion

Organic acids constitute a market with a sustained increase at present. Many of them are produced on a large scale by chemical synthesis from petroleum derivatives. Still, more recently, other alternatives, cheap and renewable sources of raw materials, are being intensively studied, among which is whey. This trend will be reinforced soon, which, together with the improvement of microbial processes, will allow more and more bioprocesses to appear at the large scale, which will become the trend of this market in the future. Among the countries whose territories contain the majority of the companies dedicated to supplying the world demand for organic acids, the People's Republic of China stands out, which is expected to continue to be the country that will dominate this market in the coming years.

Acknowledgements

The authors wish to express their gratitude to the authorities of the Universidad Técnica del Norte (UTN, Ibarra, Imbabura, Ecuador) for their unconditional support, and to Dr. Bolívar Batallas and Dr. Hernán Cadenas, Dean and Vice-Dean of their faculty, respectively, for their support for this publication.

Author details

José Manuel Pais-Chanfrau[1*], Jimmy Núñez-Pérez[1], Rosario del Carmen Espin-Valladares[1], Marcos Vinicio Lara-Fiallos[1] and Luis Enrique Trujillo-Toledo[2]

1 North-Technical University, Universidad Técnica del Norte, UTN, FICAYA, Ibarra, Imbabura, Ecuador

2 University of the Armed Forces, Universidad de las Fuerzas Armadas, ESPE, Quito, Pichincha, Ecuador

*Address all correspondence to: jmpais@utn.edu.ec

References

[1] Cherrington CA, Hinton M, Mead GC, Chopra I. Organic acids: Chemistry, antibacterial activity and practical applications. Advances in Microbial Physiology. 1991;**32**:87-108. DOI: 10.1016/S0065-2911(08)60006-5

[2] Bruice PY, editor. Organic Chemistry. 8th ed. Santa Bárbara: Pearson Education Inc.; 2016. pp. 1344

[3] Panda SK, Sahu L, Behera SK, Ray RC. Research and production of organic acids and industrial potential. In: Molina G, Kumar Gupta V, Singh BN, Gathergood N, editors. Bioprocessing for Biomolecules Production. NY: John Wiley & Sons Ltd.; 2020. pp. 195-209. DOI: 10.1002/9781119434436.ch9

[4] Capellán-Pérez I, Mediavilla M, de Castro C, Carpintero Ó, Miguel LJ. Fossil fuel depletion and socio-economic scenarios: An integrated approach. Energy. 2014;**77**:641-666

[5] Gaurav N, Sivasankari S, Kiran GS, Ninawe A, Selvin J. Utilization of bioresources for sustainable biofuels: A review. Renewable and Sustainable Energy Reviews. 2017;**73**:205-214. DOI: 10.1016/j.rser.2017.01.070

[6] Huerta Blanca BE, Díaz Yolotli Azucena T, Trinidad Adriana L. Utilización de residuos agroindustriales. Revista Sistemas Ambientales. 2008;**2**:44-50

[7] Madeira JV, Contesini FJ, Calzado F, et al. Agro-industrial residues and microbial enzymes: An overview on the eco-friendly bioconversion into high value-added products. In: Brahmachari G, editor. Biotechnology of Microbial Enzymes: Production, Biocatalysis and Industrial Applications. 1st ed. NY: Academic Press, Elsevier; 2017. pp. 475-511. DOI: 10.1016/B978-0-12-803725-6.00018-2

[8] Pandey A, Soccol CR, Nigam P, Soccol VT. Biotechnological potential of agro-industrial residues. I: Sugarcane bagasse. Bioresource Technology. 2000;**74**:69-80. DOI: 10.1016/S0960-8524(99)00142-X

[9] Wang XQ, Wang QH, Ma HZ, Yin W. Lactic acid fermentation of food waste using integrated glucoamylase production. Journal of Chemical Technology and Biotechnology. 2009;**84**:139-143. DOI: 10.1002/jctb.2007

[10] Chisti Y. Fermentation (industrial): Basic considerations. In: Batt CA, Tortorello ML, editors. Encyclopedia of Food Microbiology. 2nd ed. NY: Academic Press; 2014. pp. 751-761. DOI: 10.1016/B978-0-12-384730-0.00106-3

[11] Lizardi-Jiménez MA, Hernández-Martínez R. Solid state fermentation (SSF): Diversity of applications to valorize waste and biomass. 3 Biotech. 2017. DOI: 10.1007/s13205-017-0692-y

[12] Vandenberghe LPS, Karp SG, de Oliveira PZ, de Carvalho JC, Rodrigues C, Soccol CR. Solid-state fermentation for the production of organic acids. In: Pandey A, Larroche C, Soccol CR, editors. Current Developments in Biotechnology and Bioengineering. London: Elsevier B.V; 2018. pp. 415-434. DOI: 10.1016/B978-0-444-63990-5.00018-9

[13] Dezam APG, Vasconcellos VM, Lacava PT, Farinas CS. Microbial production of organic acids by endophytic fungi. Biocatalysis and Agricultural Biotechnology. 2017;**11**:282-287. DOI: 10.1016/j.bcab.2017.08.001

[14] Kim NJ, Lim SJ, Chang HN. Volatile fatty acid platform: Concept and application. In: Chang HN, Lee SY, Nielsen J, Stephanopoulos G, editors.

Emerging Areas in Bioengineering. Weinheim: Wiley-VCH Verlag GmbH & Co. KGaA; 2018. pp. 173-190. DOI: 10.1002/9783527803293.ch10

[15] Yin X, Li J, Shin HD, Du G, Liu L, Chen J. Metabolic engineering in the biotechnological production of organic acids in the tricarboxylic acid cycle of microorganisms: Advances and prospects. Biotechnology Advances. 2015;**36**:830-841

[16] Baumann I, Westermann P. Microbial production of short chain fatty acids from lignocellulosic biomass: Current processes and market. BioMed Research International. 2016;**2016**:15

[17] Sheldon RA. Green and sustainable manufacture of chemicals from biomass: State of the art. Green Chemistry. 2014;**16**:950-963

[18] Naraian R, Kumari S. Microbial production of organic acids. In: Gupta VK, Treichel H, Shapaval V(O), de Oliveira LA, Tuohy MG, editors. Microbial Functional Foods and Nutraceuticals. NY: John Wiley & Sons Ltd.; 2018. pp. 93-121. DOI: 10.1002/9781119048961.ch5

[19] Zumalacárregui de Cárdenas L, Zumalacárregui de Cárdenas B. Production of organic acids via fermentation of sugars generated from lignocellulosic biomass. In: Ingle AP, Chandel AK, da Silva SS, editors. Lignocellulosic Biorefining Technologies. NY: Wiley-Blackwell; 2020. pp. 203-246

[20] Bhatia SK, Yang YH. Microbial production of volatile fatty acids: Current status and future perspectives. Reviews in Environmental Science and Biotechnology. 2017;**16**:327-345

[21] Ricke SC. Perspectives on the use of organic acids and short chain fatty acids as antimicrobials. Poultry Science. 2003;**82**:632-639. DOI: 10.1093/ps/82.4.632

[22] Theron MM, Lues JFR. Organic Acids and Food Preservation. 1st ed. London: CRC Press; 2019. pp. 40. DOI: 10.1201/9781420078435

[23] Schönfeld P, Wojtczak L. Short- and medium-chain fatty acids in energy metabolism: The cellular perspective. Journal of Lipid Research. 2016;**57**:943-954

[24] Hijova E, Chmelarova A. Short chain fatty acids and colonic health. Bratislavské Lekárske Listy. 2007;**108**:254-258

[25] Panda SK, Mishra S, Kayitesi E, Ray R. Microbial-processing of fruit and vegetable wastes for production of vital enzymes and organic acids: Biotechnology and scopes. Environmental Research. 2016;**146**:161-172. DOI: 10.1016/j.envres.2015.12.035

[26] Singh R, Mittal A, Kumar M, Mehta PK. Organic acids: An overview on microbial production. International Journal of Advanced Biotechnology and Research. 2017;**8**:104-111

[27] Sauer M, Porro D, Mattanovich D, Branduardi P. Microbial production of organic acids: Expanding the markets. Trends in Biotechnology. 2008;**26**:100-108

[28] Singh Nee Nigam P. Production of organic acids from agro-industrial residues. In: Singh Nee Nigam P, Pandey A, editors. Biotechnology for Agro-Industrial Residues Utilisation: Utilisation of Agro-Residues. London: Springer Science B.V; 2009. pp. 37-60. DOI: 10.1007/978-1-4020-9942-7_3

[29] Moeller L, Strehlitz B, Aurich A, Zehnsdort A, Bley T. Optimization of citric acid production from glucose by

Yarrowia lipolytica. Engineering in Life Sciences. 2007;7:504-511

[30] Hesham AE-L, Mostafa YS, AlSharqi LEO. Optimization of citric acid production by immobilized cells of novel yeast isolates. Mycobiology. 2020;**48**:122-132. DOI: 10.1080/12298093.2020.1726854

[31] Dörsam S, Fesseler J, Gorte O, Hahn T, Zibek S, Syldatk C, et al. Sustainable carbon sources for microbial organic acid production with filamentous fungi. Biotechnology for Biofuels. 2017;**10**:1-12

[32] Shendurse AM, Khedkar CD. Lactose. In: Caballero B, Finglas PM, Toldrá F, editors. Encyclopedia of Food and Health. Oxford: Academic Press; 2015. pp. 509-516. DOI: 10.1016/B978-0-12-384947-2.00415-3

[33] Bassette R, Acosta JS. Composition of milk products. In: Wong NP, Jenness R, Keeney M, Marth EH, editors. Fundamentals of Dairy Chemistry. 3rd ed. Maryland: Aspen Publishers, Inc.; 1999. pp. 39-79. DOI: 10.1007/978-1-4615-7050-9_2

[34] Wong SY, Hartel RW. Crystallization in lactose refining—A review. Journal of Food Science. 2014;**79**:R257-R272. DOI: 10.1111/1750-3841.12349

[35] Bylund G, Malmgren B, Holanowski A, Hellman M, Mattsson G, Svensson B, et al. Dairy Processing Handbook. 3rd ed. Lund: Tetra Pak International S. A.; 2015. 486 p

[36] Ostojic S, Pavlovic M, Zivic M, Filipovic Z, Gorjanovic S, Hranisavljevic S, et al. Processing of whey from dairy industry waste. Environmental Chemistry Letters. 2005;**3**:29-32. DOI: 10.1007/s10311-005-0108-9

[37] Valta K, Damala P, Angeli E, Antonopoulou G, Malamis D, Haralambous KJ. Current treatment technologies of cheese whey and wastewater by greek cheese manufacturing units and potential valorisation opportunities. Waste and Biomass Valorization. 2017;**8**:1649-1663. DOI: 10.1007/s12649-017-9862-8

[38] Jelen P. Whey processing utilization and products. In: Fuquay JW, editor. Encyclopedia of Dairy Sciences. 2nd ed. NY: Academic Press; 2011. pp. 731-737. DOI: 10.1016/B978-0-12-374407-4.00495-7

[39] Ramchandran L, Vasiljevic T, Processing W. In: Tamime AY, editor. Membrane Processing: Dairy and Beverage Applications. NY: Blackwell Publishing Ltd; 2012. pp. 193-207. DOI: 10.1002/9781118457009.ch9

[40] Božanić R, Barukčić I, Lisak K. Possibilities of whey utilisation. Journal of Nutrition & Food Sciences. 2014;**2**(7):1036

[41] Espina VS, Jaffrin MY, Frappart M, Ding LH. Separation of casein micelles from whey proteins by high shear microfiltration of skim milk using rotating ceramic membranes and organic membranes in a rotating disk module. Journal of Membrane Science. 2008;**325**:872-879

[42] de Carvalho AF, Maubois JL. Applications of membrane technologies in the dairy industry. In: dos Reis Coimbra JS, Teixeira JA, editors. Engineering Aspects of Milk and Dairy Products. Boca Raton: Taylor and Francis Group, LLC; 2010. pp. 33-56. DOI: 10.1201/b18319-22

[43] Pouliot Y. Membrane processes in dairy technology—From a simple idea to worldwide panacea. International Dairy Journal. 2008;**18**:735-740. DOI: 10.1016/j.idairyj.2008.03.005

[44] Tamime AY, editor. Membrane Processing: Dairy and Beverage

Applications. 1st ed. Oxford: Wiley; 2013. pp. 369. DOI: 10.1002/9781118457009

[45] Vuataz G. The phase diagram of milk: A new tool for optimising the drying process. Lait. 2002;**82**: 485-500. DOI: 10.1051/lait:2002026

[46] Písecký J. Spray drying in the cheese industry. International Dairy Journal. 2005;**15**:531-536. DOI: 10.1016/j.idairyj.2004.11.010

[47] Chegini G, Taheri M. Whey powder: Process technology and physical properties: A review. Middle-East Journal of Scientific Research. 2013;**13**:1377-1387. DOI: 10.5829/idosi.mejsr.2013.13.10.1239

[48] Kelly P. Manufacture of whey protein products: Concentrates, isolate, whey protein fractions and microparticulated. In: Deeth HC, Bansal N, editors. Whey Proteins: From Milk to Medicine. London: Elsevier; 2019. pp. 97-122. DOI: 10.1016/B978-0-12-812124-5.00003-5

[49] Rossi S. CLAL.it—Italian Dairy Economic Consulting firm [Internet]. 2020. Available from: https://www.clal.it/en/?section=chisiamo

[50] OECD/FAO. OECD-FAO Agricultural Outlook 2019-2028. 2019

[51] El-Tanboly E. Recovery of cheese whey, a by-product from the dairy industry for use as an animal feed. Journal of Nutritional Health & Food Engineering. 2017;**6**:148-154

[52] Spălăţelu (Vicol) C. Biotechnological valorisation of whey. Innovative Romanian Food Biotechnology. 2012;**10**:1-8

[53] Ryan MP, Walsh G. The biotechnological potential of whey. Reviews in Environmental Science and Biotechnology. 2016;**15**:479-498

[54] Parra-Huertas R. Whey: Importance in the food industry. Revista Facultad Nacional de Agronomía Medellín. 2009;**62**:4967-4982

[55] Lappa IK, Papadaki A, Kachrimanidou V. Cheese whey processing: Integrated biorefinery. Food. 2019;**8**:347

[56] Mollea C, Marmo L, Bosco F. Valorisation of cheese whey, a by-product from the dairy industry. In: Muzzalupo I, editor. Food Industry. 1st ed. London: IntechOpen; 2013. pp. 549-588. DOI: 10.5772/53159

[57] Pais-Chanfrau JM, Núñez Pérez J, Lara Fiallos MV, et al. Milk whey—From a problematic byproduct to a source of valuable products for health and industry: An overview from biotechnology. La Prensa Medica Argentina. 2017;**103**:1-11. DOI: 10.4172/lpma.1000257

[58] Prazeres AR, Carvalho F, Rivas J. Cheese whey management: A review. Journal of Environmental Management. 2012;**110**:48-68

[59] Grosu L, Fernandez B, Grigoraş CG, Patriciu OI, Grig ICA, Nicuţă D, et al. Valorization of whey from dairy industry for agricultural use as fertiliser: Effects on plant germination and growth. Environmental Engineering and Management Journal. 2012;**11**:2203-2210

[60] Geiger B, Nguyen HM, Wenig S, Nguyen HA, Lorenz C, Kittl R, et al. From by-product to valuable components: Efficient enzymatic conversion of lactose in whey using β-galactosidase from *Streptococcus thermophilus*. Biochemical Engineering Journal. 2016;**116**:45-53

[61] Lappa IK, Papadaki A, Kachrimanidou V, Terpou A, Koulougliotis D, Eriotou E, et al. Cheese whey processing: Integrated biorefinery

concepts and emerging food applications. Food. 2019;**8**:1-37

[62] Mawson AJ. Bioconversions for whey utilization and waste abatement. Bioresource Technology. 1994;**47**:195-203

[63] Djekic I, Miocinovic J, Tomasevic I, Smigic N, Tomic N. Environmental life-cycle assessment of various dairy products. Journal of Cleaner Production. 2014;**68**:64-72. DOI: 10.1016/j.jclepro.2013.12.054

[64] Marwaha SS, Kennedy JF. Whey—Pollution problem and potential utilization. International Journal of Food Science and Technology. 1988;**23**:323-336

[65] Das B, Sarkar S, Sarkar A, Bhattacharjee S, Bhattacharjee C. Recovery of whey proteins and lactose from dairy waste: A step towards green waste management. Process Safety and Environment Protection. 2016;**101**:27-33

[66] Zadow JG. Lactose: Properties and uses. Journal of Dairy Science. 1984;**67**:2654-2679. DOI: 10.3168/jds.S0022-0302(84)81625-2

[67] Chen XY, Gänzle MG. Lactose and lactose-derived oligosaccharides: More than prebiotics? International Dairy Journal. 2017;**67**:61-72. DOI: 10.1016/j.idairyj.2016.10.001

[68] Guerrero C, Illanes A. Enzymatic production of other lactose-derived prebiotic candidates. In: Illanes A, Guerrero C, Vera C, Wilson L, Conejeros R, Scott F, editors. Lactose-Derived Prebiotics: A Process Perspective. 1st ed. Oxford: Academic Press; 2016. pp. 229-259. DOI: 10.1016/B978-0-12-802724-0.00006-8

[69] Thelwall LAW. Lactose: Chemical derivatives. In: Fox PF, editor. Advanced Dairy Chemistry. Volume 3: Lactose, Water, Salts and Vitamins. 2nd ed. London: Chapman & Hall; 1997. pp. 39-76. DOI: 10.1007/978-1-4757-4409-5_2

[70] Soccol CR, Vandenberghe LPS, Rodrigues C, et al. Production of organic acids by solid-state fermentation. In: Pandey A, Soccol CR, Larroche C, editors. Current Developments in Solid-State Fermentation. 1st ed. NY: Springer; 2008. pp. 205-229. DOI: 10.1007/978-0-387-75213-6_10

[71] Danner H, Braun R. Biotechnology for the production of commodity chemicals from biomass. Chemical Society Reviews. 1999;**28**:395-405

[72] Jiang L, Fu H, Yang HK, Xu W, Wang J, Yang ST. Butyric acid: Applications and recent advances in its bioproduction. Biotechnology Advances. 2018;**36**:2101-2117

[73] Organic acids market worth 11.39 billion USD by 2022 [Internet]. 2017. Available from: https://www.marketsandmarkets.com/PressReleases/organic-acid.asp

[74] Papagianni M. Organic acids. In: Moo-Young M, editor. Comprehensive Biotechnology. 3rd ed. Oxford: Pergamon, Elsevier B.V; 2019. pp. 85-97. DOI: 10.1016/B978-0-444-64046-8.00009-4

[75] Pappenberger G, Hohmann HP. Industrial production of L-ascorbic acid (vitamin C) and D-isoascorbic acid. Advances in Biochemical Engineering/Biotechnology. 2013;**143**:143-188. DOI: 10.1007/10_2013_243

[76] Feng XJ, Zhang HB, Liu HZ, et al. Recovery processes of organic acids from fermentation broths in the biomass-based industry. Journal of Microbiology and Biotechnology. 2016;**26**:1-8

[77] Aurich A, Specht R, Müller RA, et al. Microbiologically produced

carboxylic acids used as building blocks in organic synthesis. In: Wang X, Chen J, Quinn PJ, editors. Reprogramming Microbial Metabolic Pathways. Dordrecht: Springer Science; 2012. pp. 391-423

[78] Harrison RG, Todd PW, Rudge SR, Petrides D. Bioseparations Science and Engineering. 2nd ed. Oxford: Oxford University Press; 2015. pp. 577

[79] Bekatorou A, Dima A, Tsafrakidou P, Boura K, et al. Downstream extraction process development for recovery of organic acids from a fermentation broth. Bioresource Technology. 2016;**220**:34-37

[80] Kumar S, Babu BV. Process intensification for separation of carboxylic acids from fermentation broths using reactive extraction. I-manager's Journal on Future Engineering and Technology. 2008;**3**: 21-28. DOI: 10.26634/jfet.3.3.643

[81] Oliveira FS, Araújo JM, Ferreira R, Rebelo LPN, Marrucho IM. Extraction of l-lactic, l-malic, and succinic acids using phosphonium-based ionic liquids. Separation and Purification Technology. 2012;**85**:137-146

[82] Mamlouk D, Gullo M. Acetic acid bacteria: Physiology and carbon sources oxidation. Indian Journal of Microbiology. 2013;**53**:377-384

[83] Cheung H, Tanke RS, Torrence GP. Acetic acid. In: WILEY-VCH, editor. Ullmann's Encyclopedia of Industrial Chemistry. 7th ed. Vol. 1. Weinheim: Wiley-VCH Verlag; 2012. pp. 209-237. DOI: 10.1002/14356007.a01_045.pub2

[84] Global acetic acid market to reach 20.3 million tons by 2024. 2019. Available from: https://www.marketwatch.com/press-release/global-acetic-acid-market-to-reach-203-million-tons-by-2024-2019-10-14

[85] Acetic acid market to reach USD 21.65 billion by 2027. Reports And Data. 2020. Available from: https://www.globenewswire.com/news-release/2020/01/07/1967588/0/en/Acetic-Acid-Market-To-Reach-USD-21-65-Billion-By-2027-Reports-And-Data.html

[86] Sugisawa T, Miyazaki T, Hoshino T. Microbial production of L-ascorbic acid from D-sorbitol, L-sorbose, L-gulose, and L-sorbosone by *Ketogulonicigenium vulgare* DSM 4025. Bioscience, Biotechnology, and Biochemistry. 2005;**69**:659-662

[87] Foyer CH. Ascorbic acid. In: Alscher RG, Hess JL, editors. Antioxidants in Higher Plants. 1st ed. Boca Raton: CRC Press; 2017. pp. 28. DOI: 10.1201/9781315149899

[88] Du LD, Kong XY, Du GH. Vitamin C. In: Du GH, editor. Natural Small Molecule Drugs from Plants. Beijing: Springer; 2018. DOI: 10.1007/978-981-10-8022-7_105

[89] Magiorkinis E, Beloukas A, Diamantis A. Scurvy: Past, present and future. European Journal of Internal Medicine. 2011;**22**:147-152

[90] Danehy JP. Synthesis of ascorbic acid from lactose. 1981. US Patent 4,259,443

[91] Gelsky J. Vitamin C prices triple in a year's time. Food Bus News. [Internet] 2017. Available from: https://www.foodbusinessnews.net/articles/10836-vitamin-c-prices-triple-in-a-year-s-time [Accessed: 02 May 2020]

[92] Dwidar M, Park JY, Mitchell RJ, Sang BI. The future of butyric acid in industry. Scientific World Journal. 2012;**2012**:1-10

[93] Sjöblom M, Matsakas L, et al. Catalytic upgrading of butyric acid towards fine chemicals and biofuels. FEMS Microbiology Letters. 2016;**363**:1-7. DOI: 10.1093/femsle/fnw064

[94] Xu Z, Jiang L. Butyric acid. In: Moo-Young M, editor. Comprehensive Biotechnology. 3rd ed. Oxford: Elsevier B.V; 2019. pp. 235-243. DOI: 10.1016/B978-0-444-64046-8.00162-2

[95] Xu Q, Song L, Zhang L, Hu G, et al. Synthesis of cellulose acetate propionate and cellulose acetate butyrate in a CO_2/DMSO system. Cellulose. 2018;**25**:205-216. DOI: 10.1007/s10570-017-1539-8

[96] Dange PN, Sharma A, Rathod VK. Synthesis of methyl butyrate using heterogeneous catalyst: Kinetic studies. Catalysis Letters. 2014;**144**:1537-1546

[97] Brändle J, Domig KJ, Kneifel W. Relevance and analysis of butyric acid producing clostridia in milk and cheese. Food Control. 2016;**67**:96-113

[98] Wang J, Lin M, Xu M, Yang ST. Anaerobic fermentation for production of carboxylic acids as bulk chemicals from renewable biomass. Advances in Biochemical Engineering/Biotechnology. 2016;**156**:323-361. DOI: 10.1007/10_2015_5009

[99] Jha AK, Li J, Yuan Y, Baral N, Ai B. A review on bio-butyric acid production and its optimization. International Journal of Agriculture and Biology. 2014;**16**:1019-1024

[100] Søltoft-Jensen J, Hansen F. New chemical and biochemical hurdles. In: Sun DW, editor. Emerging Technologies for Food Processing. 1st ed. Oxford: Academic Press; 2005. pp. 387-416. DOI: 10.1016/B978-012676757-5/50017-7

[101] Show PL, Oladele KO, Siew QY, Aziz Zakry FA, Lan JCW, Ling TC. Overview of citric acid production from *Aspergillus niger*. Frontiers in Life Science. 2015;**8**:271-283

[102] Ciriminna R, Meneguzzo F, Delisi R, Pagliaro M. Citric acid: Emerging applications of key biotechnology industrial product. Chemistry Central Journal. 2017;**11**:1-9. DOI: 10.1186/s13065-017-0251-y

[103] Kristiansen B, Linden J, Mattey M. Citric Acid Biotechnology. 1st ed. London: CDC Press; 2002. pp. 190

[104] Kirimura K, Yoshioka I. Citric acid. In: Moo-Young M, editor. Comprehensive Biotechnology. 3rd ed. Oxford: Pergamon, Elsevier B.V; 2019. pp. 158-165. DOI: 10.1016/B978-0-444-64046-8.00157-9

[105] Vandenberghe LPS, Rodrigues C, de Carvalho JC, Medeiros ABP, Soccol CR. Production and application of citric acid. In: Pandey A, Soccol S, Negi CR, editors. Current Developments in Biotechnology and Bioengineering: Production, Isolation and Purification of Industrial Products. 1st ed. Oxford: Elsevier B.V; 2016. pp. 557-575. DOI: 10.1016/B978-0-444-63662-1.00025-7

[106] Alonso S, Rendueles M, Díaz M. Microbial production of specialty organic acids from renewable and waste materials. Critical Reviews in Biotechnology. 2015;**35**:497-513

[107] Ahmadi N, Khosravi-Darani K, Mortazavian AM. An overview of biotechnological production of propionic acid: From upstream to downstream processes. Electronic Journal of Biotechnology. 2017;**28**:67-75

[108] Weissermel K, Arpe HJ. Industrial Organic Chemistry. 3rd ed. Berlin: VCH; 2003. pp. 481

[109] Lidén G. Carboxylic acid production. Fermentation. 2017;**3**: 46-48. DOI: 10.3390/fermentation3030046

[110] González-García RA, McCubbin T, Navone L, Stowers C, Nielsen LK, Marcellin E. Microbial propionic acid

production. Fermentation. 2017;**3**:21-41. DOI: 10.3390/fermentation3020021

[111] Gupta A, Srivastava AK. Continuous propionic acid production from cheese whey using *in situ* spin filter. Biotechnology and Bioprocess Engineering. 2001;**6**:1-5

[112] Leroy F, De Vuyst L. Lactic acid bacteria as functional starter cultures for the food fermentation industry. Trends in Food Science and Technology. 2004;**15**:67-78

[113] Endres HJ, Siebert-Raths A. Engineering Biopolymers: Markets, Manufacturing, Properties and Applications. 1st ed. Munich: Hanser; 2011. pp. 675. DOI: 10.3139/9783446430020.fm

[114] Benninga H. A History of Lactic Acid Making: A Chapter in the History of Biotechnology. Netherlands: Springer; 1990

[115] König H, Fröhlich J. Lactic acid bacteria. In: König H, Unden G, Fröhlich J, editors. Biology of Microorganisms on Grapes, in Must and in Wine. 1st ed. Munich: Wiley; 2017. pp. 3-41. DOI: 10.1007/978-3-319-60021-5_1

[116] Castillo Martinez FA, Balciunas EM, Salgado JM, Domínguez González JM, Converti A, Oliveira RPS. Lactic acid properties, applications and production: A review. Trends in Food Science and Technology. 2013;**30**:70-83. DOI: 10.1016/j.tifs.2012.11.007

[117] Ray RC, Joshi VK. Fermented foods: Past, present and future. In: Luque A, Hegedus S, editors. Microorganisms and Fermentation of Traditional Foods. 1st ed. London: Taylor & Francis; 2014. pp. 1-36

[118] Hassan SED, Abdel-Rahman MA, Roushdy MM, Azab MS, Gaber MA. Effective biorefinery approach for lactic acid production based on co-fermentation of mixed organic wastes by *Enterococcus durans* BP130. Biocatalysis and Agricultural Biotechnology. 2019;**20**:1-9. DOI: 10.1016/j.bcab.2019.101203

[119] Chen GQ, editor. Plastics from Bacteria: Natural Functions and Applications. 1st ed. Berlin, Heidelberg: Springer-Verlag; 2010. 450 p. DOI: 10.1007/978-3-642-03287-5

[120] Kumar S, Thakur K. Bioplastics—Classification, production and their potential food applications. Journal of Hill Agriculture. 2017;**8**:118-129. DOI: 10.5958/2230-7338.2017.00024.6

[121] Petersmann S, Spoerk M, et al. Mechanical properties of polymeric implant materials produced by extrusion-based additive manufacturing. Journal of the Mechanical Behavior of Biomedical Materials. 2020;**104**:1-13. DOI: 10.1016/j.jmbbm.2019.103611

[122] Chen Q, Mangadlao JD, Wallat J, De Leon A, Pokorski JK, Advincula RC. 3D printing biocompatible polyurethane/poly(lactic acid)/graphene oxide nanocomposites: Anisotropic properties. ACS Applied Materials & Interfaces. 2017;**9**:4015-4023. DOI: 10.1021/acsami.6b11793

[123] Luiz-Ferreira RT, Cardoso-Amatte I, Assis-Dutra T, Bürger D. Experimental characterization and micrography of 3D printed PLA and PLA reinforced with short carbon fibers. Composites. Part B, Engineering. 2017;**124**:88-100. DOI: 10.1016/j.compositesb.2017.05.013

[124] Safin RR, Talipova GA, Galyavetdinov NR. Design of packaging materials based on polylactide and wood filler. International Journal of Engineering and Technology. 2018;**7**:1089-1091. DOI: 10.14419/ijet.v7i4.36.25036

[125] Cao Z, Pan H, Chen Y, Han L, Bian J, Zhang H, et al. Ductile and biodegradable poly (lactic acid) matrix film with layered structure. International Journal of Biological Macromolecules. 2019;**137**:1141-1152. DOI: 10.1016/j.ijbiomac.2019.07.047

[126] Karamanlioglu M, Preziosi R, Robson GD. Abiotic and biotic environmental degradation of the bioplastic polymer poly(lactic acid): A review. Polymer Degradation and Stability. 2017;**137**:122-130. DOI: 10.1016/j.polymdegradstab.2017.01.009

[127] Nghiem NP, Kleff S, Schwegmann S. Succinic acid: Technology development and commercialization. Fermentation. 2017;**3**:1-14

[128] Liu YP, Zheng P, Sun ZH, Ni Y, Dong JJ, Zhu LL. Economical succinic acid production from cane molasses by *Actinobacillus succinogenes*. Bioresource Technology. 2008;**99**:1736-1742. DOI: 10.1016/j.biortech.2007.03.044

[129] Shen N, Qin Y, Wang Q, Liao S, Zhu J, Zhu Q, et al. Production of succinic acid from sugarcane molasses supplemented with a mixture of corn steep liquor powder and peanut meal as nitrogen sources by *Actinobacillus succinogenes*. Letters in Applied Microbiology. 2015;**60**:544-551

[130] Okino S, Noburyu R, Suda M, Jojima T, Inui M, Yukawa H. An efficient succinic acid production process in a metabolically engineered *Corynebacterium glutamicum* strain. Applied Microbiology and Biotechnology. 2008;**81**:459-464

[131] Thakker C, Martínez I, San KY, Bennett GN. Succinate production in *Escherichia coli*. Biotechnology Journal. 2012;7:213-224

[132] Otero JM, Cimini D, Patil KR, Poulsen SG, Olsson L, Nielsen J. Industrial systems biology of *Saccharomyces cerevisiae* enables novel succinic acid cell factory. PLoS One. 2013;**8**:541-544

[133] Nattrass L, Aylott M, Higson A. NNFCC renewable chemicals factsheet: Succinic acid. 2013. Available from: http://www.nnfcc.co.uk/publications/nnfcc-renewable-chemicals-factsheet-succinic-acid

[134] Succinic acid market 2018 to 2025: Global industry analysis by product types, price, key trends & companies, opportunities. 2019. Available from: https://www.marketwatch.com/press-release/succinic-acid-market-2018-to-2025-global-industry-analysis-by-product-types-price-key-trends-companies-opportunities-2019-02-19

[135] Steiger MG, Blumhoff ML, Mattanovich D, Sauer M. Biochemistry of microbial itaconic acid production. Frontiers in Microbiology. 2013;**4**:1-5

[136] Willke T, Vorlop KD. Biotechnological production of itaconic acid. Applied Microbiology and Biotechnology. 2001;**56**:289-295

[137] Boruta T, Bizukojc M. Production of lovastatin and itaconic acid by *Aspergillus terreus*: A comparative perspective. World Journal of Microbiology and Biotechnology. 2017;**33**:33-45

[138] Roa Engel CA, Straathof AJJ, Zijlmans TW, et al. Fumaric acid production by fermentation. Applied Microbiology and Biotechnology. 2008;**78**:379-389

[139] Zhang B, Skory CD, Yang ST. Metabolic engineering of *Rhizopus oryzae*: Effects of overexpressing pyc and pepc genes on fumaric acid biosynthesis from glucose. Metabolic Engineering. 2012;**14**:512-520

[140] Talabardon M, Schwitzguébel JP, Péringer P. *Anaerobic thermophilic* fermentation for acetic acid production from milk permeate. Journal of Biotechnology. 2000;**76**:83-92

[141] Talabardon M, Schwitzguébel JP, Péringer P, Yang ST. Acetic acid production from lactose by an anaerobic thermophilic coculture immobilized in a fibrous-bed bioreactor. Biotechnology Progress. 2000;**16**:1008-1017

[142] Nayak J, Pal P. Transforming waste cheese-whey into acetic acid through a continuous membrane-integrated hybrid process. Industrial and Engineering Chemistry Research. 2013;**52**:2977-2984

[143] Pal P, Nayak J. Development and analysis of a sustainable technology in manufacturing acetic acid and whey protein from waste cheese whey. Journal of Cleaner Production. 2016;**112**:59-70

[144] Pal P, Nayak J. Acetic acid production and purification: Critical review towards process intensification. Separation and Purification Reviews. 2017;**46**:44-61

[145] Teles JC, Stolle EM, Koloda SA, Barana AC. Production of propionic acid by *Propionibacterium acidipropionici* from agroindustrial effluents. Brazilian Archives of Biology and Technology. 2019;**62**:1-12

[146] Pandey A, Srivastava S, Rai P, Duke M. Cheese whey to biohydrogen and useful organic acids: A non-pathogenic microbial treatment by *Lactobacillus acidophilus*. Scientific Reports. 2019;**9**:1-9

[147] O'Brien DJ, Panzer CC, Eisele WP. Biological production of acrylic acid from cheese whey by resting cells of *Clostridium propionicum*. Biotechnology Progress. 1990;**6**:237-242

[148] Rosa JCC, Colombo LT, Alvim MCT, Avonce N, Van Dijck P, Passos FML. Metabolic engineering of *Kluyveromyces lactis* for L-ascorbic acid (vitamin C) biosynthesis. Microbial Cell Factories. 2013;**12**(59):1-13 DOI: 10.1186/1475-2859-12-59

[149] Sauer M, Branduardi P, Valli M, Porro D. Production of L-ascorbic acid by metabolically engineered *Saccharomyces cerevisiae* and *Zygosaccharomyces bailii*. Applied and Environmental Microbiology. 2004;**70**:6086-6091

[150] Jiang L, Cui H, Zhu L, Hu Y, Xu X, Li S, et al. Enhanced propionic acid production from whey lactose with immobilized *Propionibacterium acidipropionici* and the role of trehalose synthesis in acid tolerance. Green Chemistry. 2015;**17**:250-259

[151] Vidra A, Tóth AJ, Németh Á. Complex whey utilization: The propionic acid alternative. Liquid Waste Recovery. 2017;**2**:9-12

[152] Panesar PS, Kennedy JF, Knill CJ, Kosseva M. Production of L(+) lactic acid using *Lactobacillus casei* from whey. Brazilian Archives of Biology and Technology. 2010;**53**:219-226

[153] Bernardo MP, Coelho LF, Sass DC, Contiero J. L-(+)-lactic acid production by *Lactobacillus rhamnosus* B103 from dairy industry waste. Brazilian Journal of Microbiology. 2016;**47**:640-646

[154] Luongo V, Policastro G, Ghimire A, Pirozzi F, Fabbricino M. Repeated-batch fermentation of cheese whey for semi-continuous lactic acid production using mixed cultures at uncontrolled pH. Sustainability. 2019;**11**(3330):1-12. DOI: 10.3390/su11123330

[155] Alam S, Stevens D, Bajpai R. Production of butyric acid by batch fermentation of cheese whey with *Clostridium beijerinckii*. Journal of Industrial Microbiology. 1988;**2**:359-364

[156] Vandák D, Tomáška M, Zigová J, Šturdík E. Effect of growth supplements and whey pretreatment on butyric acid production by *Clostridium butyricum*. World Journal of Microbiology and Biotechnology. 1995;**11**:363

[157] Lee PC, Lee WG, Kwon S, Lee SY, Chang HN. Batch and continuous cultivation of *Anaerobiospirillum succiniciproducens* for the production of succinic acid from whey. Applied Microbiology and Biotechnology. 2000;**54**:23-27

[158] Wan C, Li Y, Shahbazi A, Xiu S. Succinic acid production from cheese whey using *Actinobacillus succinogenes* 130 Z. Applied Biochemistry and Biotechnology. 2008;**145**:111-119

[159] Podlesny M, Wyrostek J, Kucharska J, et al. A new strategy for effective succinic acid production by *Enterobacter sp*. LU1 using a medium based on crude glycerol and whey permeate. Molecules. 2019;**24**:4543

[160] Kaur R, Kaur R, Sharma A, et al. Microbial production of dicarboxylic acids from edible plants and milk using GC-MS. Journal of Analytical Science and Technology. 2018;**9**:1-10

[161] Alonso S, Rendueles M, Díaz M. Simultaneous production of lactobionic and gluconic acid in cheese whey/glucose co-fermentation by *Pseudomonas taetrolens*. Bioresource Technology. 2015;**196**:314-323

[162] El-Holi MA, Al-Delaimy KS. Citric acid production from whey with sugars and additives by *Aspergillus niger*. African Journal of Biotechnology. 2003;**2**:356-359

[163] Amado IR, Vázquez JA, Pastrana L, Teixeira JA. Cheese whey: A cost-effective alternative for hyaluronic acid production by *Streptococcus zooepidemicus*. Food Chemistry. 2016;**198**:54-61

[164] Sheng J, Ling P, Wang F. Constructing a recombinant hyaluronic acid biosynthesis operon and producing food-grade hyaluronic acid in *Lactococcus lactis*. Journal of Industrial Microbiology & Biotechnology. 2014;**42**:197-206

[165] Beccerra M, Díaz Prado S, Rodríguez-Belmonte E, Cerdán ME, González Siso MI. Metabolic engineering for direct lactose utilization by *Saccharomyces cerevisiae*. Biotechnology Letters. 2002;**24**:1391-1396

Chapter 6

Value-Added Compounds with Health Benefits Produced from Cheese Whey Lactose

Hada María Guevara-Alvarado, Néstor Gutiérrez-Méndez, Esther Carrillo-Pérez, Einar Vargas-Bello-Pérez and José Carlos Rodríguez-Figueroa

Abstract

Cheese whey (CW) is the yellow-green liquid main by-product from cheese manufacturing. Historically, it has been recognized as a major environmental pollutant. Nowadays, it represents a source of high-quality nutrients, such as lactose. Enzymatic bioprocesses, chemical synthetic reactions and microbial bioprocesses use lactose as substrate to obtain relevant derivatives such as lactitol, lactulose, lactosucrose, sialyllactose, kefiran and galacto-oligosaccharides. These lactose derivatives stimulate the growth of indigenous bifidobacteria and lactobacilli improving the intestinal motility, enhancing immunity and promoting the synthesis of vitamins. Also, they have versatile applications in pharmaceutical, biotechnological and food industries. Therefore, this book chapter shows the state of the art focusing on recent uses of CW lactose to produce value-added functional compounds and discusses new insights associated with their human health-promoting effects and well-being.

Keywords: cheese whey, bioprocesses, value-added functional compounds, lactose, kefiran

1. Introduction

Cheese whey (CW) is the yellow-green liquid main by-product from the manufacture of cheese [1]. "Serum milk" remaining after the precipitation and removal of milk proteins by proteolytic enzymes or acid may also be defined as CW [2]. Industrial cheese manufacturing processes produce sweet or acid whey (**Figure 1**). Normally, the production of 1 kg of cheese requires 10 kg of milk originating 9 kg of CW [3, 4]. Worldwide cheese production was estimated by FAO (Food and Agricultural Organization) at 22.65 M tons in 2014 [5]. Therefore, CW is estimated at 203.9 M tons. Besides, the global growth rate of it is parallel to the cheese production and it has been calculated about 2% per annum [6]. This amount represents a challenge difficult to deal with.

Previous studies have reported high quantity of organic matter in CW [7]. The chemical composition of it is shown in **Table 1**. This by-product has 50, 000–102, 000 mg/L Chemical Oxygen Demand (COD) and 27, 000–60, 000 mg/L Biological

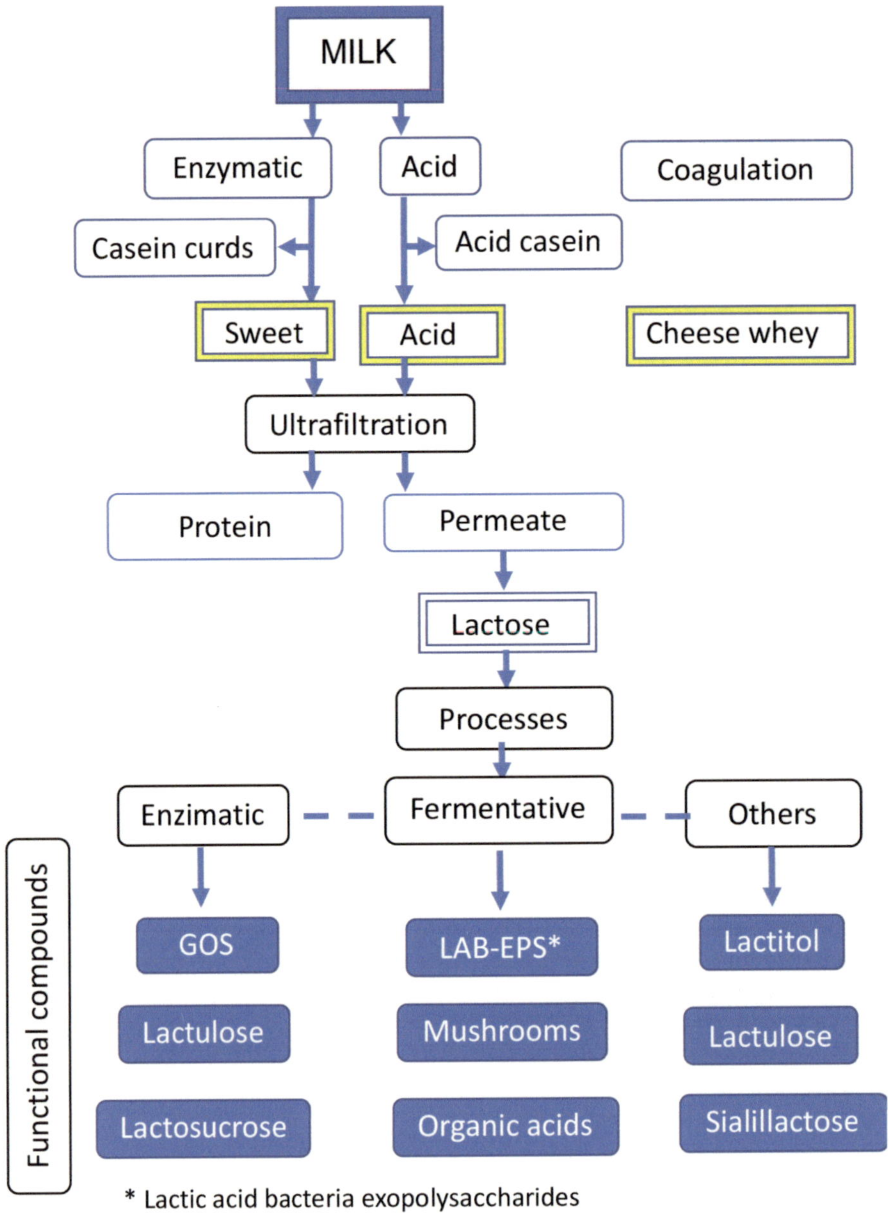

Figure 1.
Overview of value-added functional compounds using lactose from cheese whey as substrate.

	Chemical composition (g/100 g)				
Cheese whey type	**Total solids**	**Lactose**	**Protein**	**Fat**	**Ash**
Sweet cheese whey	6.7	4.8	0.6	0.25	0.54
Sweet cheese whey permeate	5.5	4.7	0.05	<0.01	0.51
Acid cheese whey	5.1	4.4	0.73	0.05	0.6
Acid cheese whey pemeate	5.8	4.3	0.06	<0.01	0.56

Table 1.
Chemical composition of different types of cheese whey.

Oxygen Demand (BOD) [7]. Due to its high BOD, CW presents 175-fold higher organic load than typical sewage effluents [6]. Lactose (4-*O*-ß-D-galactopyranosyl-D-glucose) is one of the main CW components. It causes about 90% of COD and BOD [8]. Moreover, CW represents about 85–95% of the milk volume. This amount has been partially land spreading, disposal to natural water bodies or municipal sewer systems [2–4]. Consequently, it is considered a major pollutant to the environment.

On the other hand, CW is a liquid with high nutritional content [9]. It retains about 55% of the milk nutrients such as lactose, whey proteins, lipids, vitamins and minerals (**Table 1**). The chemical make-up of it can vary depending on the animal species from which milk was obtained [1]. It has been reported that about 50% of the total CW worldwide production is used it [10]. Animal feeding or as an ingredient in therapeutic formulations and food applications are common CW uses [11, 12]. Several technological approaches have been developed to transform it into value-added compounds reducing the environmental impact. CW processing is carried out directly by physical or thermal treatments to obtain protein isolate (WPI), whey protein concentrate (WPC), whey protein hydrolysates, whey permeate, lactose and other fractions. Indirectly, CW is used as substrate for enzymatic/microbial bioprocesses to produce biogas, bioethanol, bioprotein, biopolymers, flavors and organic acids among others [11, 13]. Thus, it is an excellent substrate for physical treatments, enzymatic catalysis and metabolic microbial reactions that could be exploited by the medical, agri-food and biotechnology industries [6].

CW is a source of functional proteins, peptides, lipids and carbohydrates. This by-product is the main source of lactose manufactured on an industrial scale, as well as a low-cost substrate able to reduce high production costs. This disaccharide ($C_{12}H_{22}O_{11}$) is the most abundant in CW representing around 70–72% (w/w) of the total solids [14]. The manufacture of edible lactose includes physical treatments such as ultrafiltration, nanofiltration, concentration, crystallization, washed and dried [15]. Lactose is a valuable ingredient used in a wide variety of products, such as bread, supplement in baby milk formulae, confectionaries and excipients for pharmaceutical products [16].

The use of lactose as raw material is a key point to several industrial and laboratory transformation processes. This disaccharide represents an ideal substrate to obtain relevant lactose derivatives associated with health-promoting benefits. For example, lactulose ($C_{12}H_{22}O_{11}$; 4-*O*-ß-D-galactopyranosyl-ß- D-fructofuranose), which is synthetized by chemical isomerization, is a typical prebiotic as bifidus factor added to infant formulae [16].

Enzymatic catalytic and fermentation bioprocesses also use CW lactose as substrate to produce important value-added functional compounds [16, 17]. Kefiran, organic acids, lactosucrose and galacto-oligosaccharides (GOS) are some of the most representative compounds able to improve human health and well-being (**Figure 1**). For instance, GOS [Gal-(Gal)n-Glu] that are produced by enzymatic polymerization using ß-galactosidase, improve gut health. Besides, exopolysaccharides such as kefiran are synthetized by the fermenting bioprocesses of lactic acid bacteria [16, 18]. Therefore, this chapter discusses recent uses of lactose from CW to produce value-added functional compounds. New insights associated with human health benefits of these compounds are explored.

2. Lactose derivatives with health benefits

2.1 Lactitol

Lactitol is a lactose-derived compound defined as synthetic sugar alcohol ($C_{12}H_{24}O_{11}$; 4-O-ß-D-galactopyranosyl-D-glucitol; molecular weight (MW),

344.31 g/mol) comprising galactose (D-Gal) and D-sorbitol. This compound is synthetized by catalytic hydrogenation reactions using lactose as substrate. Industrially, this chemical process is based on the addition of molecular hydrogen to the carbonyl group of the glucose molecules. This chemical reaction needs 110°-150°C temperature, 20–70 bars of hydrogen gas pressure, as well as 1.5–10% Ni, Pd or Ru transition metals in either carbon or alumina. Lactitol is the primary reaction product with reaction yields of >90%. This polyol has been used in the food industry as a relevant ingredient in desserts, bakery products, chewing gums, chocolate and confectionary products. One of the advantages of lactitol used as sweetener is that it can be metabolized by saccharolytic bacteria providing only 2 kcal/g. It also exerts properties such as cryoprotectant, dryoprotectant, stabilizer agent, hydrogel delivering bioactive compounds and additive for the development of biosensors [19].

Several human health benefits have been associated to lactitol intake. Clinical trials have demonstrated positive gastrointestinal health benefits of this polyol [20]. A random-effect meta-analysis of lactitol supplementation on adult constipation demonstrated favorable efficacy and tolerance when it was compared to stimulant laxatives and placebo. Lactitol was able to induce increased fecal volume by stimulating peristalsis [21]. In fact, lactitol is one of the most frequently prescribed osmotic laxative agents to treat constipation [22]. Investigations performed on the effectiveness of lactitol for treatment of several types of hepatic encephalopathy in infants, children and elderly subjects have demonstrated positive results. Actually, lactitol is recommended as a first-line treatment for hepatic encephalopathy as a result of decreasing the absorption and production of ammonia and reducing the intestinal pH [23, 24]. Indeed, in the last years advances in the field of nanomedicine had led to the development of polylactitol as a multifunctional carrier for liver cancer therapy [24].

Lactitol is a non-digestible carbohydrate with prebiotic effect. Prebiotic is a substrate that is selectively utilized by host microorganisms conferring a health benefit [25]. Previous studies have reported relevant lactitol symbiotic effects. Medical practitioners frequently recommend them as therapeutics. Recently, it was demonstrated that the consumption of the symbiotic combination with this lactose-derived prebiotic, *Bifidobacterium bifidum* and *Lactobacillus acidophilus* was able to eradicate OXA-48-producing *Enterobacteriaceae*. The measure of this metabolite is used as prophylaxis to prevent intestinal translocations in neutropenic patients and for the prevention of pneumonia [26]. Also, the symbiotic supplementation on the gut microbiota of healthy elderly volunteers with *Lactobacillus acidophilus* NCFM and lactitol improved their health status modifying the intestinal environment and the microbiota composition. It was observed an increasing lactobacilli and bifidobacteria and a possible stabilizing effect on *Blautia coccoides-Eubacterium rectale* and *Clostridium cluster* XIV levels [27]. Even though the Federal and Drug Administration (FDA) agency categorized lactose-derived prebiotic as "GRAS" (Generally Recognized as Safe), the excess consumption has adverse effects such as osmotic diarrhea, abdominal pain and vomiting [20]. It has been reported that the maximum permissive dosage of lactitol for Japanese adults not to induce transitory diarrhea was 0.36 g/kg of body weight [28]. It was also found that the dose of this lactose-derived prebiotic treatment is age- and case-dependent [20].

2.2 Lactulose

Lactulose is a semi-synthetic disaccharide ($C_{12}H_{22}O_{11}$; 4-O-ß-D-galactopyranosyl-ß-D-fructofuranose; MW, 342.30 g/mol) comprising D-galactose (D-Gal) and D-fructose (D-Fru) linked by ß-1-4 glycosidic bond [16, 29].

Commercially, this artificial disaccharide is synthetized by alkaline isomerization of lactose via the Lobry de Bruyn e Alberda van Ekenstein rearrangement in which the D-glucose unit at the reducing end of the lactose molecule is converted to D-fructose. The maximum yield of lactulose relative to initial lactose concentration adding complexing agents may reach up to 88% [30]. Electro-activation is a novel eco-friendly technology able to synthetize lactulose from CW-lactose at maximum yield of 35% [31]. In the last decade lactulose has been synthetized at lab scale using lactose as substrate by the transgalactosylation activity of ß-glucosidase [29]. This disaccharide is formed in milk during heat treatments also, so pasteurized milk usually has <100 mg/L lactulose content, meanwhile ultra-high temperature (UHT) milk generally has a lactulose content over 500 mg/L [32]. Actually, this polyol has demonstrated versatile applications in pharmaceutical and food industries. Lactulose can be found as relevant functional ingredient of infant food formulae, fermented dairy products, bakery products, confectionary products and soy milk [33, 34].

Previous studies have reported remarkable health benefits associated to lactulose consumption. In fact, this disaccharide is used in clinical practice since 1957. This disaccharide is lactose-derived prebiotic able to prevent and to treat diseases [35]. Lactulose is only metabolized by specific species of colonic microbiota through ß-glucosidase activity altering the microbial balance by increasing the probiotic growth and reducing putrefactive bacteria. Consequently, lowering intestinal pH, enhanced colonic motility, reduced concentration of ammonia and improved absorption of minerals are also benefits of the physiological action of lactulose upon bacterial metabolism in the large intestine [30].

In silico, *in vitro* and *in vivo* studies have demonstrated the efficacy of lactulose in the treatment of several diseases [35]. Since the 1960s, patients of all ages have been prescribed with lactulose to treat constipation, even if it is chronic. This lactose-derived prebiotic is an osmotic laxative [33]. The effect of lactulose was studied in healthy volunteers. A significant increase of *Bifidocaterium*, *Lactobacillus* and *Streptococcus* was reached, meanwhile the population of coliforms, *Bacteroides*, *Clostridium* and *Eubacterium* was significantly decreased. These changes in the microbiota reduced activity of pro-carcinogenic enzymes, increased short-chain fatty acids in feces and pH decreased [36]. Clinical trials also have reported favorable results using lactulose to treat hepatic encephalopathy and chronic kidney disease [30]. Recently, the prebiotic effect of lactitol, raffinose, oligofructose and lactulose was evaluated on *Lactobacillus* spp. and bacterial vaginosis-associated organisms (BV) and *Candida albicans*. Results showed that lactulose had the most broadly and specifically growth stimulation on vaginal lactobacilli and did not to stimulate BV or *Candida albicans* [37]. On the other side, *in vitro* and *in vivo* studies have confirmed that lactulose possesses patient- and dose-dependent prebiotic properties [35].

2.3 Sialyllactose

Sialyllactose ($C_{23}H_{39}NO_{19}$; NeuAcα2-xD-galactopyranosyl-α-D-glucopyranoside; MW, 633.6 g/mol) is essentially sialic acid (N-acetylneuraminic acid, NeuAc) bound to a lactose molecule. This lactose-derived compound is naturally found in high concentrations at the beginning of lactation in colostrum and decreases towards the end of lactation [16]. The predominant forms of sialyllactose are 6′-sialyllactose and 3′-sialyllactose. The concentrations of 6′-sialyllactose in human colostrum is 250–1300 mg/L, meanwhile the concentration of 3′-sialyllactose in bovine colostrum is 354–1250 mg/L [38]. These lactose-derived compounds are extracted from CW using ultra and nanofiltration processes on a tangential flow type laboratory scale membrane filtration system [39]. Even though this still an expensive procedure to extract sialyllactose, some infant formulae use it as functional ingredient [16, 40].

In vivo studies have demonstrated the ability of sialyllactose to improve positively in health. Pathogenic microorganisms have been effectively inhibited using it [16, 40]. It was reported that the consumption of dietary sialyllactose modified the colonic microbiota, e.g. Bacteroidetes were significantly increased, meanwhile Firmicutes and Cyanobacteria were significantly decreased. Moreover, this lactose-derived prebiotic was able to diminish stressor-induced alterations in colonic mucosa and anxiety-like behavior [41]. One of the major causes of morbidity and mortality in premature infants is necrotizing enterocolitis (NEC). Recently, it was found that human milk oligosaccharides 2′-fucosyllactose and 6′-sialyllactose can reduce NEC and attenuate NEC inflammation [42]. In addition, intact sialylated oligosaccharides can be absorbed in concentrations high enough to modulate the immunological system and facilitate proper brain development during infancy [43].

3. Functional compounds bio-produced using lactose as substrate

3.1 Biocatalytic processes

3.1.1 Galacto-oligosaccharides

Galacto-oligosaccharides [Gal-(Gal)n-Glu] are lactose-derived non-digestable oligosaccharides (GOS) recognized as relevant functional compounds. Industrially, GOS are produced using CW lactose as substrate through biocatalytic reaction. Lactose is transgalactosylated by ß-galactosidases enzymes (E.C. 3.2.1.23) from several microbial strains [44]. GOS are the best substitute for human oligosaccharides, have a sweet taste, low energy value (2 kcal/g), as well as tolerate high temperatures and low pHs. So, they are widely used in the food industry as functional ingredient in the manufacturing of infant formula, confectionary, chewing gum, yogurt, ice cream and bakery products [45].

Previous studies have demonstrated the impact of GOS promoting gut health and well-being. These lactose-derived prebiotics serve as substrates for the microbiota, improve saccharolytic metabolic activities and stimulate the growth of indigenous bifidobacteria and lactobacilli. In consequence, the formation of volatile fatty acids, lowering of the luminal pH and decreased formation of toxic secondary bile acids are microbial metabolic associated effects. Also, they inhibit the formation of toxic bacterial metabolites, such as ammonia, hydrogen disulphide, phenolic compounds and biogenic amines [44]. Moreover, GOS have a bifidus factor similar to the effect of human milk oligosaccharides stimulating the growth of specific intestinal microbiota, improving the intestinal motility, enhancing immunity, promoting the synthesis of vitamins, reducing the high levels of cholesterol and triglycerides and decreasing the risk of colon cancer development [45, 46].

3.1.2 Lactosucrose

Lactosucrose is an oligosaccharide comprising Gal, Fru and Glu. This carbohydrate molecule ($C_{18}H_{38}O_{16}$; MW, 510.4 g/mol) is a ß-D-fructofuranosyl-4-O- ß-D-galactopyranosyl-α-D-glucopyranoside [16]. Lactosucrose can be regarded as a condensate of sucrose and galactose molecules or lactose and fructose molecules. Production protocols include transferring the ß -galactosyl group produced by the decomposition of lactose to the C4 hydroxyl group of glucosyl in sucrose by the enzymatic activity of ß-galactosidase (E.C. 3.2.1.23). Also, it can be produced by the catalysis of ß-fructofuranosidase (E.C. 3.2.1.26) or levansurase (E.C. 2.4.1.10) transferring the fructose group generated by the decomposition of sucrose to the C1 hydroxyl group at the reducing end of the lactose. Industrially, ß-fructofuranosidase is one of the most common enzymes used to the production of lactosucrose due to its availability and low cost [47]. This non-reducing trisaccharide is an ingredient of cosmetic and pharmaceutical products. Moreover, it is widely used in a large number of functional foods. In fact, in Japan, lactosucrose has the status of FOSHU ingredient. So, it has been used in a large number of healthy foods and drinks, such as bakery products, yogurt, ice creams, infant formula, chocolates, juice and mineral water [48].

In the last decades, the demand for lactosucrose has significantly increased. This can be explained by the widely uses of it in the preparation of functional foods. Lactosucrose is well known by its prebiotic effect. *In vivo* studies in animals, as well as in humans have demonstrated the association between lactosucrose consumption and health-promoting effects. Their review includes enhancement of beneficial bacteria and or inhibition of pathogenic microorganisms, decrease of fecal pH, production of short chain fatty acids and gases, reduction of putrefactive products, enhancement of intestinal absorption of minerals, treatment of chronic inflammatory bowel diseases, normalization of intestinal microflora and prevention of abdominal symptoms of lactose intolerance [48].

3.2 Microbial bioprocesses

3.2.1 Lactic acid bacteria exopolysaccharides (LAB-EPS)

Lactic acid bacteria (LAB) play a key role in the fermentation processes of food worldwide. These group of microorganisms improve the preservation, enhance sensory characteristics, increase nutritional values of a large variety of food and beverages products and have been recognized by their health-promoting attributes [49]. Several LAB have the ability to produce exopolysaccharides (EPS) as cell wall constituents named peptidoglycan located in the extracellular medium without covalent bounds with bacterial membrane [49, 50]. EPS are a diverse group of high-molecular-mass polysaccharides in terms of chemical composition, quantity, molecular size, charge, presence of side chains rigidity of the molecules, including mechanisms of synthesis [49, 51].

LAB-EPS are classified depending on the composition of the main chain and their mechanisms of synthesis. They can be divided into homopolysaccharides (HoPs) or heteropolysaccharides (HePs) In general, HoPs contain only one type of monosaccharide (glucose or fructose) through linear or branched α or β links, with more than 10^6 Da molecular mass. These EPS are produced in grams per liter by *Lactobacillus*, *Leuconosctoc*, *Oenococcus* and *Weissella* extracellularly from sucrose or starch without noncarbohydrate groups. On the other side, HePs contain more than one type of monosaccharide, mainly glucose, galactose and rhamnose together through α and β links, typically branched with 10^4–10^6 Da molecular mass. Most of them are produced in milligrams per liter by *Lactobacillus*, *Lactococcus*, *Bifidobacterium* and *Streptococcus* from intracellular intermediates with the presence of noncarbohydrates groups [51].

Kefiran is the main HePs synthetized by kefir grains microorganisms. Kefir grains are a consortium of symbiotic LAB, acetic acid bacteria, bifidobacteria and yeast microorganisms embedded in a matrix of proteins, lipids, polysaccharides and water [52] . These microorganisms are able to synthetize kefiran from CW lactose even if it is deproteinized [53]. In fact, using CW lactose as a fermentation medium presents the opportunity to create value-added products [54]. *Lactobacillus kefiranofaciens* has been identified as the most important kefiran producer. Previous study demonstrated this extracellular polysaccharide is water soluble and it has the same amounts of D-glucose and D-galactose, approximately. Kefiran has several relevant applications within the biotechnology, food and pharmaceutical industries [52]. Therefore, increasing attention has been paid to these EPS.

Kefiran is a natural EPS that offers relevant food and pharmaceutical industrial advantages. It could be added to a formulation or it could be produced *in situ* through fermentation processes. As a polymer, kefiran exert versatile functionality. In food industry, for example it has widely applications such as stabilizer, additive, film-forming agent and gelling agent. In recent years, it has been discovered novel nano applications of this HePs, e.g. kefiran-based bio-nanocomposites and kefiran based nanofibers. Moreover, this bio-molecule also has shown biological activity properties. Several *in vitro* and *in vivo* studies have demonstrated the ability of kefiran to increase peritoneal IgA, reduce blood pressure induced hypertension, wound healing, antioxidant activity, antitumoral activity, favor the activity of peritoneal macrophages, modulation of the intestinal immune system and protection of epithelial cells against, prevent several cancer, anti-inflammatory and prebiotic effect [55, 56].

HoPs have also potential uses in the food and pharmaceutical industries. Fructans (levan and inulin-like), α -glucans (dextran, reuteran, alternan and mutan) and β -glucans are the most important HoPs [49, 51]. HoPs such as dextran have been using in bakery products improving softness or in confectionary, ice cream, frozen and dried-food and non-alcoholic wort-based beverages as stabilser. Levan and inulin-like HoPs can be used as fat substitute and sugar replacer, respectively. Besides, these HoPs may influence human host health. For example, β-glucans have demonstrated a cholesterol-lowering effect increasing cardiovascular health. Moreover, *Lactobacillus delbrueckii* subsp. *bulgaricus* strains HoPs removed cholesterol from *in vitro* culture media. Indeed, HoPs have been recognized by their benefits on the microbial gut modulation acting as prebiotics [51].

3.2.2 Mushrooms

In recent years, the use of CW for mycelial growth has been explored. CW as substrate offers a wide diversity of nutrients such as proteins, carbohydrates, lipids, vitamins and minerals. On the other side, the metabolism of mycelia of fungi produced edible mushrooms utilizes the nutrients from the medium to bioaccumulate microelements such as Se, Fe and Zn. Therefore, the use of CW for mycelial growth may be a valuable nutritional supplement, reducing the impact of discharging CW to the environment and biofortifies mushrooms composition.

The nutritional, culinary and nutraceutical properties of mushrooms have attracted the researchers, pharmacists and nutritionists attention. The chemical composition of mushrooms includes bioactive molecules such as polysaccharides, terpenoids, low molecular weight proteins, glycoproteins among others that play a key role in boosting immune strength, lowering risks of cancers, inhibiting of tumoral growth, maintaining of blood sugar, etc. [57].

Information on mushrooms chemical composition, nutritional value and therapeutic properties has expanded during the last few years. *Pleurotus* spp. (oyster mushrooms) are one of the most cultivated mushrooms worldwide [58]. Recently, it was demonstrated that the mycelial growth of *Pleurotus djamor* in a liquid culture medium containing CW was able to produce bioactive compounds such as ergosterol and β-glucans. The addition of selenium to the medium decreased the concentration of lactose. Moreover, it was observed that the mycelium showed potential in absorbing and accumulating elements e.g. Ca, Fe, Mg, K and Zn from the CW [59].

3.2.3 Organic acids

Several organic acids are produced during the metabolic pathways of the fermentation processes. Some organic acids e. g. lactic acid, propionic acid, butyric acid, isobutyric acid, acetic acid, capric acid, caproic acid, caprylic acid, lactobionic acid, etc., are responsible for characteristic flavors [60, 61]. However, they play a key role as functional compounds enhancing health-promoting effects and well-being. It has been demonstrated that conjugated linoleic acid (CLA, 9,11-Octadecadienoic acid, MW, 280.4 g/mol) modulate the fatty acid composition of the liver and adipose tissue of the host [62]. Indeed, succinic acid ($C_4H_6O_4$, MW, 118.09 g/mol) has shown its ability to stabilize the hypoxia and cellular stress conditions focusing on the maintenance of homeostasis in aging hypothalamus. Therefore, it is hypothesized that succinate has the potential to restore the loss in functions associated with cellular senescence and systematic aging [63]. Most of the commercial succinic acid production is done by chemical technologies like catalytic hydrogenation or electrolytic reduction of maleic anhydride. In the last years, it was found that it can be produced using CW and lactose as substrates by *Actinobacillus succinogenes* 130Z in a batch fermentation [64].

According to the international market demands, lactobionic acid, fumaric acid and glucaric acid are classified as high value-added compounds [61]. These organic acids have demonstrated potential uses in food, medicine, pharmaceutical, cosmetic and chemical industries [61, 65, 66]. Glucaric acid ($C_6H_{10}O_8$, MW, 210.14 g/mol) is found in vegetables and fruits, mainly grapefruits, apples, oranges and cruciferous vegetables. Commercially, it is synthetized by chemical oxidation of glucose releasing toxic byproducts. Thus, microbial fermentation of glucose by *Saccharomyces cerevisiae* and *Escherichia coli* has been proposed as alternative. This organic acid and its derivatives increases detoxification of carcinogens compounds and tumor promoters [67, 68]. Fumaric acid (trans-1,2-ethylenedicarboxylic acid, MW, 116.07 g/mol) is traditionally synthetized from maleic anhydride, which in turn is produced from butane. Nowadays, the production of this organic acid may be done by fermenting glucose through the metabolic pathways of *Rhizopus* species, also fixing CO_2. Fumaric acid is widely used as starting material for polymerization and esterification reactions to produce paper and unsaturated polyester resins. In medicine field, it can be used to treat psoriasis, meanwhile it is also used as food and beverage additive. Moreover, Fumaric acid supplements have the ability to reduce methane emissions of cattle [66].

Lactobionic acid (4-O-ß -galactopyranosyl-D-gluconic acid, MW, 358.3 g/mol) is a high value-added lactose derivative. This organic acid has received growing attention due to its multiple applications in cosmetics, chemical, pharmaceutical, biomedicine, and food industries [61]. Lactobionic acid production is based on chemical synthesis requiring high amounts of energy and costly metal catalysts [69]. Nowadays, lactobionic acid is able to be bio-produced either through enzymatic or microbial biosynthesis at cost-effective and environmentally friendly using cheese whey lactose. In fact, high-level production of it has been recently reported controlling pH and temperature during the fermentation of lactose with *Pseudomonas taetrolens* [70]. Lactobionic acid offers wide versatile uses in nanotechnology, tissue engineering and drug-delivery systems, antibiotics, preservative solutions for organ transplantation, anti-aging, regenerative skin-care, sugar-based surfactant. Also, this value-added compound functions as food additive, gelling agent, solubilizing agent, sweetener, water holding capacity agent and bioactive ingredient enhancing calcium absorption, antioxidant activity and exerting prebiotic effects [61].

Lactic acid (2-hydroxipropionic acid, MW 90.08 g/mol) is an organic acid with a prime position due to its versatile applications in textile, leather, chemical, pharmaceutical and food industries. Lactic acid applications associated to food and food-related represent 85% of total production, approximately. This organic acid has been recognized as GRAS by the FDA [71]. It is used as flavoring, buffering agent, inhibitor of bacterial spoilage, acidulant, dough conditioner and emulsifier [72]. Most of lactic acid is produced through microbial fermentation, mainly *Lactobacillus delbrueckii or Lactobacillus amylophilus* strains, using beet extracts, molasses, starchy and cellulosic materials and cheese whey [71].

Polylactic acid is a biocompatible polymer with unique properties. Lactic acid and lactide are the building blocks to obtain it through a polycondensation reaction. This biodegradable and renewable biopolymer is a relevant alternative to plastics derived from petrochemicals, so its demand has been increasing considerably. In fact, the global polylactic acid market was expected to grow over 1.2 million tons in 2020. Nowadays, most polylactic acid is manufactured for single-use applications in packaging, including food packaging supplies [73]. However, it has important biomedical uses, due to its GRAS status recognized by the FDA. This biomaterial has been transformed into sutures, scaffolds, cell carriers and drug delivery systems such as liposomes, polymeric nanoparticles, dendrimers and micelles [74, 75].

4. Conclusions

Cheese whey production is increasing worldwide every year. Even though CW is considered a major environmental pollutant, due to its quantity and quality of chemical components, there is a huge opportunity to use it as raw material to produce value-added functional compounds. CW lactose is an excellent substrate to obtain high quality products able to improve human health and well-being, e.g. lactitol, lactosucrose, GOS, lactulose, sialyllactose and organic acids. For example, GOS and sialyllactose have a bifidus factor similar to the effect of human milk oligosaccharides stimulating the growth of specific intestinal microbiota, enhancing immunity, promoting the synthesis of vitamins and decreasing the risk of colon cancer. Moreover, microbial bioprocesses use CW lactose to produce relevant health-promoting metabolites such as kefiran and organic acids. Future perspectives are focusing on the sustainable transformation of CW lactose as by product into value-added functional compounds to be used as novel ingredients in a diverse formulation of food, pharmaceutical, and cosmetic new products. Therefore, additional research concentrated on the development of innovative technological processes, more efficient and able to discover new bioactive compounds are essential.

Acknowledgements

Authors thank the Mexican Science and Technology Council (CONACYT) for financing the publication of this book chapter through the project 258483.

Author details

Hada María Guevara-Alvarado[1], Néstor Gutiérrez-Méndez[2], Esther Carrillo-Pérez[3], Einar Vargas-Bello-Pérez[4] and José Carlos Rodríguez-Figueroa[3]*

1 Escuela Nacional Toribio Rodríguez de Mendoza, Perú

2 Universidad Autónoma de Chihuahua, México

3 Universidad de Sonora, México

4 University of Copenhagen, Denmark

*Address all correspondence to: jose.rodriguez@unison.mx

References

[1] Ryan MP, Walsh G. The biotechnological potential of whey. Rev Environ Sci Biotechnol 2016;15:479-98. https://doi.org/10.1007/s11157-016-9402-1.

[2] Panesar P, Kennedy J, Gandhi D, Bunko K. Bioutilisation of whey for lactic acid production. Food Chem 2007;105:1-14. https://doi.org/10.1016/j.foodchem.2007.03.035.

[3] Prazeres AR, Carvalho F, Rivas J. Cheese whey management: A review. J Environ Manage 2012;110:48-68. https://doi.org/10.1016/j.jenvman.2012.05.018.

[4] Jelen P. Dried Whey, Whey Proteins, Lactose and Lactose Derivative Products. In: Tamime A, editor. Dairy Powders Conc. Prod., Wiley-Blackwell; 2009, p. 255-67. https://doi.org/10.1002/9781444322729.ch7.

[5] Livestock Processed. Statistics Division, Food and Agriculture Organization of the United Nations n.d. http://www.fao.org/faostat/en/#data/QP/visualize (accessed May 6, 2020).

[6] Smithers GW. Whey and whey proteins—From 'gutter-to-gold.' Int Dairy J 2008;18:695-704. https://doi.org/10.1016/j.idairyj.2008.03.008.

[7] Carvalho F, Prazeres AR, Rivas J. Cheese whey wastewater: Characterization and treatment. Sci Total Environ 2013. https://doi.org/10.1016/j.scitotenv.2012.12.038.

[8] Kolev S. General Characteristics and Treatment Possibilities of Dairy Wastewater – A Review. Food Technol Biotechnol 2017;53:237-42. https://doi.org/10.17113/ft.

[9] Banaszewska A, Cruijssen F, Claassen GDH, van der Vorst JGAJ. Effect and key factors of byproducts valorization: The case of dairy industry. J Dairy Sci 2014;97:1893-908. https://doi.org/10.3168/jds.2013-7283.

[10] González Siso MI. The biotechnological utilization of cheese whey: A review. Bioresour Technol 1996;57:1-11. https://doi.org/10.1016/0960-8524(96)00036-3.

[11] Yadav JSS, Yan S, Pilli S, Kumar L, Tyagi RD, Surampalli RY. Cheese whey: A potential resource to transform into bioprotein, functional/nutritional proteins and bioactive peptides. Biotechnol Adv 2015;33:756-74. https://doi.org/10.1016/j.biotechadv.2015.07.002.

[12] El-Tanboly E. Recovery of Cheese Whey, a by-Product from the Dairy Industry for use as an Animal Feed. J Nutr Heal Food Eng 2017;6. https://doi.org/10.15406/jnhfe.2017.06.00215.

[13] Audic J-L, Chaufer B, Daufin G. Non-food applications of milk components and dairy co-products: A review. Lait 2003;83:417-38. https://doi.org/10.1051/lait:2003027ï.

[14] Jelen P. Whey Processing: Utilization and Products. In: Roginski H, editor. Encycl. Dairy Sci., Academic Press; 2002, p. 2739-45. https://doi.org/10.1016/b0-12-227235-8/00511-3.

[15] Durham RJ. Modern approaches to lactose production. Dairy-Derived Ingredients Food Nutraceutical Uses, Elsevier Ltd; 2009, p. 103-44. https://doi.org/10.1533/9781845697198.1.103.

[16] Seki N, Saito H. Lactose as a source for lactulose and other functional lactose derivatives. Int Dairy J 2012;22:110-5. https://doi.org/10.1016/j.idairyj.2011.09.016.

[17] Lappa IK, Papadaki A, Kachrimanidou V. Cheese Whey

Processing : Integrated Biorefinery. Foods 2019;8:347 (8-15). https://doi.org/10.3390/foods8080347.

[18] Hasheminya SM, Dehghannya J. Novel ultrasound-assisted extraction of kefiran biomaterial, a prebiotic exopolysaccharide, and investigation of its physicochemical, antioxidant and antimicrobial properties. Mater Chem Phys 2020;243:122645. https://doi.org/10.1016/j.matchemphys.2020.122645.

[19] Martínez-Monteagudo SI, Enteshari M, Metzger L. Lactitol: Production, properties, and applications. Trends Food Sci Technol 2019;83:181-91. https://doi.org/10.1016/j.tifs.2018.11.020.

[20] Nath A, Haktanirlar G, Varga Á, Molnár MA, Albert K, Galambos I, et al. Biological activities of lactose-derived prebiotics and symbiotic with probiotics on gastrointestinal system. Med 2018;54. https://doi.org/10.3390/medicina54020018.

[21] Miller LE, Tennilä J, Ouwehand AC. Efficacy and tolerance of lactitol supplementation for adult constipation: A systematic review and meta-analysis. Clin Exp Gastroenterol 2014;7:241-8. https://doi.org/10.2147/CEG.S58952.

[22] Prasad VGM, Abraham P. Management of chronic constipation in patients with diabetes mellitus. Indian J Gastroenterol 2017;36:11-22. https://doi.org/10.1007/s12664-016-0724-2.

[23] Gluud L, Vilstrup H, Morgan M. Encephalopathy in people with cirrhosis (Review). Cochrane Database Syst Rev 2018. https://doi.org/10.1002/14651858.CD012410.pub2.Copyright.

[24] Hong SJ, Ahn MH, Sangshetti J, Arote RB. Sugar alcohol-based polymeric gene carriers: Synthesis, properties and gene therapy applications. Acta Biomater 2019;97:105-15. https://doi.org/10.1016/j.actbio.2019.07.029.

[25] Gibson GR, Hutkins R, Sanders ME, Prescott SL, Reimer RA, Salminen SJ, et al. Expert consensus document: The International Scientific Association for Probiotics and Prebiotics (ISAPP) consensus statement on the definition and scope of prebiotics. Nat Rev Gastroenterol Hepatol 2017;14:491-502. https://doi.org/10.1038/nrgastro.2017.75.

[26] Ramos-Ramos JC, Lázaro-Perona F, Arribas JR, García-Rodríguez J, Mingorance J, Ruiz-Carrascoso G, et al. Proof-of-concept trial of the combination of lactitol with Bifidobacterium bifidum and Lactobacillus acidophilus for the eradication of intestinal OXA-48-producing Enterobacteriaceae. Gut Pathog 2020;12:1-8. https://doi.org/10.1186/s13099-020-00354-9.

[27] Björklund M, Ouwehand AC, Forssten SD, Nikkilä J, Tiihonen K, Rautonen N, et al. Gut microbiota of healthy elderly NSAID users is selectively modified with the administration of Lactobacillus acidophilus NCFM and lactitol. Age (Omaha) 2012;34:987-99. https://doi.org/10.1007/s11357-011-9294-5.

[28] Oku T, Nakamura S, Ichinose M. Maximum permissive dosage of lactose and lactitol for transitory diarrhea, and utilizable capacity for lactose in Japanese female adults. J Nutr Sci Vitaminol (Tokyo) 2005;51:51-7. https://doi.org/10.3177/jnsv.51.51.

[29] Vera C, Guerrero C, Aburto C, Cordova A, Illanes A. Conventional and non-conventional applications of β-galactosidases. Biochim Biophys Acta - Proteins Proteomics 2020;1868:140271. https://doi.org/10.1016/j.bbapap.2019.140271.

[30] Schuster-Wolff-Bühring R, Fischer L, Hinrichs J. Production and

physiological action of the disaccharide lactulose. Int Dairy J 2010;20:731-41. https://doi.org/10.1016/j.idairyj.2010.05.004.

[31] Kareb O, Champagne CP, Aïder M. Contribution to the production of lactulose-rich whey by in situ electro-isomerization of lactose and effect on whey proteins after electro-activation as confirmed by matrix-assisted laser desorption/ionization time-of-flight-mass spectrometry and sodium . J Dairy Sci 2016;99:2552-70. https://doi.org/10.3168/jds.2015-10037.

[32] Rasooly A. Biosensor technologies. Methods 2005;37:1-3. https://doi.org/10.1016/j.ymeth.2005.05.004.

[33] Aït-Aissa A, Aïder M. Lactulose: Production and use in functional food, medical and pharmaceutical applications. Practical and critical review. Int J Food Sci Technol 2014;49:1245-53. https://doi.org/10.1111/ijfs.12465.

[34] Nooshkam M, Babazadeh A, Jooyandeh H. Lactulose: Properties, techno-functional food applications, and food grade delivery system. Trends Food Sci Technol 2018;80:23-34. https://doi.org/10.1016/j.tifs.2018.07.028.

[35] Ruszkowski J, Witkowski JM. Lactulose: Patient- and dose-dependent prebiotic properties in humans. Anaerobe 2019;59:100-6. https://doi.org/10.1016/j.anaerobe.2019.06.002.

[36] Ballongue J, Schumann C, Quignon P. Effects of lactulose and lactitol on colonic microflora and enzymatic activity. Scand J Gastroenterol Suppl 1997;32:41-4. https://doi.org/10.1080/00365521.1997.11720716.

[37] Collins SL, Mcmillan A, Seney S, Veer C Van Der, Kort R, Sumarah MW. crossm Promising Prebiotic Candidate Established by Evaluation of 2018;84:1-15.

[38] Ten Bruggencate SJ, Bovee-Oudenhoven IM, Feitsma AL, van Hoffen E, Schoterman MH. Functional role and mechanisms of sialyllactose and other sialylated milk oligosaccharides. Nutr Rev 2014;72:377-89. https://doi.org/10.1111/nure.12106.

[39] Sousa YRF, Araújo DFS, Pulido JO, Pintado MME, Martínez-Férez A, Queiroga RCRE. Composition and isolation of goat cheese whey oligosaccharides by membrane technology. Int J Biol Macromol 2019;139:57-62. https://doi.org/10.1016/j.ijbiomac.2019.07.181.

[40] Lane JA, Mariño K, Rudd PM, Carrington SD, Slattery H, Hickey RM. Methodologies for screening of bacteria-carbohydrate interactions: Anti-adhesive milk oligosaccharides as a case study. J Microbiol Methods 2012;90:53-9. https://doi.org/10.1016/j.mimet.2012.03.017.

[41] Tarr AJ, Galley JD, Fisher SE, Chichlowski M, Berg BM, Bailey MT. The prebiotics 3'Sialyllactose and 6'Sialyllactose diminish stressor-induced anxiety-like behavior and colonic microbiota alterations: Evidence for effects on the gut-brain axis. Brain Behav Immun 2015;50:166-77. https://doi.org/10.1016/j.bbi.2015.06.025.

[42] Sodhi CP, Wipf P, Yamaguchi Y, Fulton WB, Kovler M, Niño DF, et al. The human milk oligosaccharides 2′-fucosyllactose and 6′-sialyllactose protect against the development of necrotizing enterocolitis by inhibiting toll-like receptor 4 signaling. Pediatr Res 2020:1-11. https://doi.org/10.1038/s41390-020-0852-3.

[43] Lis-Kuberka J, Orczyk-Pawiłowicz M. Sialylated oligosaccharides and glycoconjugates of human milk. The impact on infant and newborn

protection, development and well-being. Nutrients 2019;11. https://doi.org/10.3390/nu11020306.

[44] Schaafsma G. Lactose and lactose derivatives as bioactive ingredients in human nutrition. Int Dairy J 2008;18:458-65. https://doi.org/10.1016/j.idairyj.2007.11.013.

[45] Roberfroid M, Gibson GR, Hoyles L, McCartney AL, Rastall R, Rowland I, et al. Prebiotic effects: Metabolic and health benefits. Br J Nutr 2010;104. https://doi.org/10.1017/S0007114510003363.

[46] Vera C, Córdova A, Aburto C, Guerrero C, Suárez S, Illanes A. Synthesis and purification of galacto-oligosaccharides: state of the art. World J Microbiol Biotechnol 2016;32. https://doi.org/10.1007/s11274-016-2159-4.

[47] Xiao Y, Chen Q, Guang C, Zhang W, Mu W. An overview on biological production of functional lactose derivatives. Appl Microbiol Biotechnol 2019:3683-91. https://doi.org/10.1007/s00253-019-09755-6.

[48] Silvério SC, Macedo EA, Teixeira JA, Rodrigues LR. Perspectives on the biotechnological production and potential applications of lactosucrose: A review. J Funct Foods 2015;19:74-90. https://doi.org/10.1016/j.jff.2015.09.014.

[49] Zannini E, Waters DM, Coffey A, Arendt EK. Production, properties, and industrial food application of lactic acid bacteria-derived exopolysaccharides. Appl Microbiol Biotechnol 2016;100:1121-35. https://doi.org/10.1007/s00253-015-7172-2.

[50] Badel S, Bernardi T, Michaud P. New perspectives for Lactobacilli exopolysaccharides. Biotechnol Adv 2011;29:54-66. https://doi.org/10.1016/j.biotechadv.2010.08.011.

[51] Lynch KM, Zannini E, Coffey A, Arendt EK. Lactic Acid Bacteria Exopolysaccharides in Foods and Beverages: Isolation, Properties, Characterization, and Health Benefits. Annu Rev Food Sci Technol 2018;9:155-76. https://doi.org/10.1146/annurev-food-030117-012537.

[52] Dailin DJ, Elsayed EA, Othman NZ, Malek R, Phin HS, Aziz R, et al. Bioprocess development for kefiran production by Lactobacillus kefiranofaciens in semi industrial scale bioreactor. Saudi J Biol Sci 2014;23:495-502. https://doi.org/10.1016/j.sjbs.2015.06.003.

[53] By P, Sebastia N, Rimada AND, Anali A, Graciela A. Polysaccharide production by kefir grains during whey fermentation. J Dairy Res # Propr J Dairy Res 2001;68:653-61. https://doi.org/10.1017\S0022029901005131.

[54] Briczinski EP, Roberts RF. Production of an exopolysaccharide-containing whey protein concentrate by fermentation of whey. J Dairy Sci 2002;85:3189-97. https://doi.org/10.3168/jds.S0022-0302(02)74407-X.

[55] Moradi Z, Kalanpour N. Kefiran, a branched polysaccharide: Preparation, properties and applications: A review. Carbohydr Polym 2019;223. https://doi.org/10.1016/j.carbpol.2019.115100.

[56] Prado MR, Bland??n LM, Vandenberghe LPS, Rodrigues C, Castro GR, Thomaz-Soccol V, et al. Milk kefir: Composition, microbial cultures, biological activities, and related products. Front Microbiol 2015. https://doi.org/10.3389/fmicb.2015.01177.

[57] Rathore H, Prasad S, Sharma S. Mushroom nutraceuticals for improved nutrition and better human health: A review. PharmaNutrition 2017;5:35-46. https://doi.org/10.1016/j.phanu.2017.02.001.

[58] Aida FMNA, Shuhaimi M, Yazid M, Maaruf AG. Mushroom as a potential source of prebiotics : a review. Trends Food Sci Technol 2009;20:567-75. https://doi.org/10.1016/j.tifs.2009.07.007.

[59] Velez MEV, da Luz JMR, da Silva M de CS, Cardoso WS, Lopes L de S, Vieira NA, et al. Production of bioactive compounds by the mycelial growth of Pleurotus djamor in whey powder enriched with selenium. Lwt 2019;114:108376. https://doi.org/10.1016/j.lwt.2019.108376.

[60] Xiang H, Sun-Waterhouse D, Waterhouse GIN, Cui C, Ruan Z. Fermentation-enabled wellness foods: A fresh perspective. Food Sci Hum Wellness 2019;8:203-43. https://doi.org/10.1016/j.fshw.2019.08.003.

[61] Alonso S, Rendueles M, Díaz M. Bio-production of lactobionic acid: Current status, applications and future prospects. Biotechnol Adv 2013;31:1275-91. https://doi.org/10.1016/j.biotechadv.2013.04.010.

[62] O'Shea EF, Cotter PD, Stanton C, Ross RP, Hill C. Production of bioactive substances by intestinal bacteria as a basis for explaining probiotic mechanisms: Bacteriocins and conjugated linoleic acid. Int J Food Microbiol 2012;152:189-205. https://doi.org/10.1016/j.ijfoodmicro.2011.05.025.

[63] Chen TT, Maevsky EI, Uchitel ML. Maintenance of homeostasis in the aging hypothalamus: The central and peripheral roles of succinate. Front Endocrinol (Lausanne) 2015;6:1-11. https://doi.org/10.3389/fendo.2015.00007.

[64] Louasté B, Eloutassi N. Succinic acid production from whey and lactose by Actinobacillus succinogenes 130Z in batch fermentation. Biotechnol Reports 2020;27:23-7. https://doi.org/10.1016/j.btre.2020.e00481.

[65] Moon TS, Yoon SH, Lanza AM, Roy-Mayhew JD, Jones Prather KL. Production of glucaric acid from a synthetic pathway in recombinant *Escherichia coli*. Appl Environ Microbiol 2009;75:589-95. https://doi.org/10.1128/AEM.00973-08.

[66] Roa Engel CA, Straathof AJJ, Zijlmans TW, Van Gulik WM, Van Der Wielen LAM. Fumaric acid production by fermentation. Appl Microbiol Biotechnol 2008;78:379-89. https://doi.org/10.1007/s00253-007-1341-x.

[67] Żółtaszek R, Hanausek M, Kiliańska ZM, Walaszek Z. Biologiczna rola kwasu D -glukarowego i jego pochodnych ; potencjalne zastosowanie w medycynie The biological role of D -glucaric acid and its derivatives : Potential use in medicine. Postep Hig Med Dosw Online 2008:451-62.

[68] Chen N, Wang J, Zhao Y, Deng Y. Metabolic engineering of *Saccharomyces cerevisiae* for efficient production of glucaric acid at high titer. Microb Cell Fact 2018;17:1-11. https://doi.org/10.1186/s12934-018-0914-y.

[69] Kuusisto J, Tokarev A V., Murzina E V., Roslund MU, Mikkola JP, Murzin DY, et al. From renewable raw materials to high value-added fine chemicals-Catalytic hydrogenation and oxidation of d-lactose. Catal Today 2007;121:92-9. https://doi.org/10.1016/j.cattod.2006.11.020.

[70] Kim JH, Jang YA, Seong SB, Jang SA, Hong SH, Song JK, et al. High-level production and high-yield recovery of lactobionic acid by the control of pH and temperature in fermentation of Pseudomonas taetrolens. Bioprocess Biosyst Eng 2020;43:937-44. https://doi.org/10.1007/s00449-020-02290-z.

[71] John RP, G.S. A, Nampoothiri KM, Pandey A. Direct lactic acid fermentation: Focus on simultaneous saccharification and lactic acid

production. Biotechnol Adv 2009;27: 145-52. https://doi.org/10.1016/j.biotechadv.2008.10.004.

[72] John RP, Nampoothiri KM, Pandey A. Fermentative production of lactic acid from biomass: An overview on process developments and future perspectives. Appl Microbiol Biotechnol 2007;74:524-34. https://doi.org/10.1007/s00253-006-0779-6.

[73] VanWouwe P, Dusselier M, Vanleeuw E, Sels B. Lactide Synthesis and Chirality Control for Polylactic acid Production. ChemSusChem 2016;9:907-21. https://doi.org/10.1002/cssc.201501695.

[74] Tyler B, Gullotti D, Mangraviti A, Utsuki T, Brem H. Polylactic acid (PLA) controlled delivery carriers for biomedical applications. Adv Drug Deliv Rev 2016;107:163-75. https://doi.org/10.1016/j.addr.2016.06.018.

[75] Singhvi MS, Zinjarde SS, Gokhale D V. Polylactic acid: synthesis and biomedical applications. J Appl Microbiol 2019;127:1612-26. https://doi.org/10.1111/jam.14290.